The Political Economy of Mechanization in U.S. Agriculture

Westview Replica Editions

The concept of Westview Replica Editions is a response to the continuing crisis in academic and informational publishing. Library budgets for books have been severely curtailed. Ever larger portions of general library budgets are being diverted from the purchase of books and used for data banks, computers, micromedia, and other methods of information retrieval. Interlibrary loan structures further reduce the edition sizes required to satisfy the needs of the scholarly community. Economic pressures on the university presses and the few private scholarly publishing companies have severely limited the capacity of the industry to properly serve the academic and research communities. As a result, many manuscripts dealing with important subjects, often representing the highest level of scholarship, are no longer economically viable publishing projects--or, if accepted for publication, are typically subject to lead times ranging from one to three years.

Westview Replica Editions are our practical solution to the problem. We accept a manuscript in camera-ready form, typed according to our specifications, and move it immediately into the production process. As always, the selection criteria include the importance of the subject, the work's contribution to scholarship, and its insight, originality of thought, and excellence of exposition. The responsibility for editing and proofreading lies with the author or sponsoring institution. We prepare chapter headings and display pages, file for copyright, and obtain Library of Congress Cataloging in Publication Data. A detailed manual contains simple instructions for preparing the final typescript, and our editorial staff is always available to answer questions.

The end result is a book printed on acid-free paper and bound in sturdy library-quality soft covers. We manufacture these books ourselves using equipment that does not require a lengthy make-ready process and that allows us to publish first editions of 300 to 600 copies and to reprint even smaller quantities as needed. Thus, we can produce Replica Editions quickly and can keep even very specialized books in print as long as there is a demand for them.

About the Book and Author

The Political Economy of Mechanization in U.S. Agriculture

Barry L. Price

For more than a decade the trend toward increased mechanization in U.S. agriculture has been the source of farm worker protests, legislative hearings, and lawsuits. (The recent case pitting the University of California's prestigious agriculture research establishment against Cesar Chavez and the United Farm Workers is a prominent example of such litigation.) A key question in the controversy is whether federal and state governments should continue to invest more than $1 billion per year in the development of large-scale, capital-intensive technologies known to have significant social costs. Opponents of continued public support for these new technologies argue that they will eliminate thousands of farm jobs when the nation already suffers from a serious unemployment problem; proponents contend that such capital-intensive technologies keep food prices down for consumers while generating the potential for increased wages for farm workers.

This book explores both sides of the debate, tracing the history of the mechanization issue and assessing the economic and sociological bases of the opposing positions. Maintaining that present methods of analysis are not adequate for resolving the conflict, Professor Price suggests an alternative approach, highlighted by a detailed case study of the costs and benefits generated by a new harvest technology adopted in the tomato-processing industry in California. He singles out the role of market structure as the most important variable in the distribution of benefits resulting from mechanization. Finally he relates his research findings to policy alternatives concerning farm mechanization in general, as well as to other problems involving technological change.

Barry L. Price is assistant professor of government at Angelo State University, San Angelo, Texas.

The Political Economy of Mechanization in U.S. Agriculture

Barry L. Price

Routledge
Taylor & Francis Group
LONDON AND NEW YORK

First published 1983 by Westview Press

Published 2019 by Routledge
52 Vanderbilt Avenue, New York, NY 10017
2 Park Square, Milton Park, Abingdon, Oxon OX14 4RN

Routledge is an imprint of the Taylor & Francis Group, an informa business

Library of Congress Cataloging in Publication Data
Price, Barry L., 1946-
The political economy of mechanization in U.S. agriculture.

(A Westview replica edition)
1. Farm mechanization--United States. 2. Agricultural innovations--United States. 3. Agricultural price supports--United States. 4. Unemployment, Technological --United States. 5. Farms, Small--United States. 6. Agriculture--Economic aspects--United States. 7. Agriculture and state--United States. I. Title.
HD1459.P74 1983 338.1'61'0973 83-10412

ISBN 13: 978-0-367-29492-2 (hbk)

For Sheldon, Roslyn and Edith

Contents

Tables and Figures

Tables

Figures

Preface

The continuing mechanization of American agriculture has become a surprisingly volatile political issue. During the past decade, it has sparked farm worker protests, legislative hearings, and headline grabbing lawsuits. Because of its controversial nature the mechanization issue also has generated a great deal of scholarly research. This research has been conducted by agricultural engineers, plant geneticists, rural sociologists, and agricultural economists to name just a few. Naturally, scholars in each of these disparate fields address the mechanization issue from the perspectives of their own academic disciplines. They also use the research methodologies common to their respective disciplines, and disseminate their findings in professional journals familiar to colleagues in their particular fields. Understandable as it is, this discipline by discipline approach makes it difficult for anyone other than dedicated professional scholars to gain a viable understanding of the mechanization issue.

The Political Economy of Mechanization in U.S. Agriculture is an attempt to change this situation. The major purpose of this book is to integrate in a single volume the best research efforts addressing the mechanization issue, regardless of the academic discipline involved. In integrating this material, I have attempted to simplify it where possible so that those other than professional scholars can use the book to gain a more sophisticated view of the mechanization process. I hope that the thousands of activists who have been involved with this issue over the years will be able to sit down with this book for an afternoon or two and come away with a better understanding of the mechanization issue than would otherwise be possible. A second major purpose of the book is to demonstrate, particularly through the case study of the electronic tomato sorter, that it is possible to increase our understanding of the mechanization process by taking a more holistic approach, even when this forces us away from the more "objective" approach typically employed in studying this issue.

Because The Political Economy of Mechanization in U.S. Agriculture is an effort to integrate the work of existing research on the mechanization issue, whatever insights it provides can be traced largely to those whose names appear in the footnotes to each chapter. But, this book also owes a great deal to many whose names aren't footnoted. During the time that I was preparing this manuscript, I had hundreds of extremely helpful conversations with both faculty and students at the University of California, Davis. For one interested in learning about American agriculture, this community is truly a national resource. Among those whose help I simply couldn't have gotten along without are: Sheldon Gellar, who provided the intellectual inspiration necessary for this book; Don Ethridge, who tirelessly tutored me in the intricacies of agricultural economics, and Alfred Diamant, who demanded the intellectual discipline necessary to complete this book. Any shortcomings that exist are, of course, ultimately my own.

Barry L. Price

1 The Political Character of Mechanized Agriculture

THE CURRENT CONTROVERSY

The mechanization of agriculture currently is one of the most prominent issues on the American political agenda. The issue is now most visible in California where it has generated numerous farmworker protests, several legislative hearings, and even a headline grabbing lawsuit. But, as massive mechanization efforts proceed in other farming areas, the controversy surrounding this process has increasingly gained national attention. [1] At the core of this controversy is a debate over whether federal and state governments should continue to pour over $1 billion per year into agricultural research aimed at developing new technologies with significant social costs. [2] The bulk of this publicly funded research is done under the auspices of the United States Department of Agriculture and the State Land Grant University system. As the primary laboratory for agricultural research in the United States, the USDA - Land Grant University Complex is in a critical position to influence the rate and kind of technological change occurring in American agriculture. It does this in the long run by formulating the problems towards which agricultural scientists direct their energies. For many years now these institutions have defined their long-term research problem as primarily one of increasing production efficiency. Having set their institutional sights on the target of production efficiency, policymakers at these institutions have strongly encouraged the fastest possible development of large-scale mechanical innovations designed to reduce farm labor, and biochemical innovations designed to increase crop yields. [3]

Opponents of continued public support for mechanization research emphasize that the large-scale, capital-intensive technologies developed by the USDA - Land Grant Complex generate serious social costs. They argue that these new technologies are eliminating thousands of harvest-period jobs at a time when the nation is already struggling with a serious unemployment problem. They also point out that the new technology is forcing thousands of small farmers out of farming every year. Unable to afford the heavy capital investment required for such technologies, and unable to compete without them, small farmers are forced to abandon farming altogether. As farmworkers and small farmers are forced off the land, the rural communities which have depended on them for sustenance also are withering away and dying. Finally, opponents of further state support for mechanization research note that the net result of this process is a food production system dominated by corporate farmers. Agribusiness giants use their political and economic muscle to sell consumers ever more adulterated and unhealthy foods at ever more inflated and administered prices. [4]

Proponents of continued public support for mechanization research reject the dismal picture painted above. They do not deny that large-scale capital-intensive farming technologies have generated some social costs. They emphasize, however, that these costs are minor when compared to the extensive benefits these innovations generate. Moreover, they argue that these benefits are enjoyed by a wide range of groups and classes in society. They point out that consumers in particular have been the primary beneficiaries of mechanization research because it is only continued technological advance which allows farmers to keep food costs down. They also emphasize that by sharply increasing productivity, the new technologies allow for increased wages and profits to farmworkers and farmers who remain on the land. Finally, they argue that mechanization in agriculture contributes to national prosperity by forcing unneeded farmworkers and small farmers off the land and into more productive urban employment. [5]

The current controversy over state support for mechanization research reflects as much disagreement over who controls the decisions behind this research as it does over who gains or loses from it. According to opponents of further public support for mechanization research, decision makers within the USDA - Land Grant University Complex have formulated their major research problem as one of increasing production efficiency, primarily because of the lobbying efforts of the agribusiness community. Having established a general technological bias in public sector research institutions, large growers and major farm machinery corporations have further exerted their influence to encourage the kind of production technologies which bring them the greatest profits regardless of the social costs involved. This view of the decision-making process behind technological change in agriculture was recently alleged in a law suit filed by the California Rural Legal Assistance Society against Regents at the University of California. Specifically, the suit charges that U.C. Regents and other state officials have supported mechanization research because they have benefited personally from it through their investments in agribusiness corporations and their ownership of California farmland. [6]

Opponents of existing mechanization research programs point out that had policymakers directing public sector agriculture research formulated their long-term research problem differently, both the rate and kind of agricultural innovations produced in these institutions may have been changed. Had they, for example, defined their primary research problem as one of stabilizing the farm population, they might have directed more research energy toward such non-technological outputs as establishing viable farm equipment cooperatives and developing innovative farm credit schemes. This focusing of research energies on farmer adjustment problems no doubt would have slowed that rate at which new farm technologies were being generated by public sector research institutes. It also might have encouraged a different kind of farm technology. In particular, it might have encouraged the development of more simple, multi-purpose farm machinery that could be used economically on smaller acreages rather than the more highly specialized, large-scale equipment that has, in fact, been developed. [7]

Supporters of present public sector research priorities reject the "pressure-group" view of the decision-making process behind the objectives outlined above. They acknowledge that the USDA and Land Grant universities have defined their primary research problem as one of increasing farm production efficiency. They argue, however, that this research strategy has emerged primarily from a concern for the public interest, rather than as a result of pressure from agribusiness. They contend that it is based entirely on the sincerely held conviction that the benefits generated by rapid technological change in agriculture far outweigh whatever social costs this process may entail. [8] As to the charge that within the general bias of production efficiency public sector research leans toward providing technologies that are particularly profitable for agribusiness, they reply that the bias of technological change is explained solely by the search for cost minimizing techniques. If land is a relatively scarce and expensive input, farmers will demand, and public sector research will generate, technologies with a land-saving bias. If labor is a relatively scarce and expensive farming input, farmers will demand, and public sector research will generate, technologies with a labor-saving bias. [9]

A HISTORICAL PERSPECTIVE

The mechanization of agriculture is not a new phenomenon in American society. When this country was founded, 95 percent of the nation's population was engaged in agriculture. Today less than 4 percent of the population earns its livelihood in farming. [10] In general, the mechanization of agriculture has been an evolutionary process, punctuated by occasional bursts of innovative activity. The first such burst occurred between 1850 and 1880 when farmers adopted the mechanical reaper and the steel plow, and substituted animal power for manpower. The second burst occurred around the turn of the twentieth century and was ushered in by the introduction of the gasoline powered tractor. During the two decades following this innovation, farmers substituted machine for animal power in a wide variety of farm operations. The most recent burst of agricultural innovation has been in full stride since the end of World War II and is still underway today. During this era, farmers have adopted large-scale harvesting equipment to replace hundreds of thousands of farm laborers in a wide array of agricultural crops. [11] By the late 1940s, a mechanical cotton harvester had been developed which could strip 90 percent of the cotton from a field at a single pass and maintain a working speed of 2.5 miles per hour. By 1951, a potato harvesting machine was available which could dig, sort and pack "without the touch of human hands." By 1952, 80 percent of California's sugar beet crop could be harvested by a machine capable of collecting as much as forty tons of beets per hour. [12] Machines were also taking over in the orchards. Lemon groves were:

> trimmed by hedging machines that maneuvered a bank of circular saws fourteen inches in diameter arranged on a metal beam that moved them around and above the trees, giving them crew cuts that left them handier for picking. The lemons themselves, after being sorted and graded by electronic eyes, could be handled and packed in bulk, eliminating expensive hand-wrapping. An electronic sorter could separate 2,200 lemons into four shades of green and pack them in boxes. [13]

More recently, growers have adopted more mechanized techniques to harvest grapes and process tomatoes, onions and tobacco. [14]

While the mechanization of agriculture has been underway in this country since at least the mid 1800s, it has only recently become a major political issue. In part, this is because mechanization has not always involved the direct substitution of machine for human power that we see today. When farmers in the nineteenth century adopted the steel plow and mechanical reaper, they accomplished more than simply reducing the amount of labor needed to produce a bushel of grain. They also used the new equipment to expand their operations into previously untilled lands. This meant that all or part of the labor saved by the new machinery might be kept on to handle the increased output made possible by mechanization. Even the early use of the gasoline powered tractor in the first decades of the twentieth century did not always involve the direct substitution of machine for human power that characterizes the mechanization process today. According to the authors of a U.S. Department of Agriculture study issued in 1941:

> Development of large-capacity tractor equipment has been the foremost cause of the extension of the small grain belt westward into semiarid sections where low yield formerly prevented crop production and the extension of cotton production into the High Plains area in Texas and Oklahoma. By far, the greater part of the increase in crop acreage from 1919 to 1929 was in these western areas...with the tractor as the outstanding source of power. [15]

By the early 1930s, the potential for new farm machinery to expand farm acreage in the United States had been largely exploited. Subsequent mechanization required direct substitution of machines for human labor and precipitated the massive displacement of farm labor that we see as the correlate of this process today.

Even massive labor displacement, however, did not initially focus farm labor protest on the mechanization process itself. During the 1930s, for example, widespread adoption of the general purpose tractor by southern cotton growers displaced at least 100,000 sharecropper families. [16] Yet, the Southern Tenant Farmers Union, the principal voice of displaced tenant farmers during this period, never made mechanization per se a major political issue. Instead, it chose to focus the nation's attention primarily on the discriminatory allocation of New Deal agricultural benefits and the oppressive treatment of tenant farmers by Southern planters. [17] From the late 1940s to the late 1960s, farmers were able to introduce many more labor-saving devices still without focusing farm labor protest on the mechanization process itself. During these years, they mechanized sugar beets, grapes, prunes, peaches, potatoes and cotton to name just a few. And, in each of these crops, mechanization involved significant labor displacement. Adoption of the mechanical cotton harvester, for example, displaced 50,000 farm laborers in California alone. [18] Yet, a review of major farm labor protests in California during these years indicates that farm laborers were still primarily concerned with their low wages and poor working conditions, rather than mechanization per se. [19]

Why did farm laborers refrain from challenging the legitimacy of mechanization even when they saw that it inevitably precipitated significant labor displacement? Part of the explanation, no doubt, can be found in the fact that the majority of workers being displaced by machines throughout most of the 1950s and 1960s were not American citizens. Most of these workers were Mexican citizens who had come to the U.S. either illegally or under the provisions of "Public Law 78." Under this law, if the Secretary of Labor certified that there was a shortage of domestic labor, he was empowered to admit Mexican laborers into the country to perform seasonal agricultural work. [20] The number of Mexican "braceros" entering the U.S. under this program ranged from 192,000 in 1951, its first year of operation, to 445,197 in 1956. [21] Roughly 75 percent of these "braceros" worked in California or Texas, where they were joined by as many as 1,000,000 illegal entrants a year. [22] Obviously, when these workers lost their jobs to the machine, they felt they had little right to complain. And even those who did feel such a right had little chance of making mechanization a viable political issue when they were not even U.S. citizens.

All of this began to change in 1964. In that year, Congress repealed Public Law 78. From that point on, mechanization in the fields began to take the jobs of American workers. And these workers increasingly were represented by a vocal and politically influential union. No longer undermined by a large pool of Mexican "braceros", Ceasar Chavez and other farm labor leaders began to make significant progress in their efforts to develop the United Farmworkers Union (UFW). By the late 1960s, they were locked in a bitter struggle with some of California's largest fruit and vegetable growers. It was during the course of this struggle that UFW leaders came face to face with the grower threat to mechanize rather than unionize. And, by this time, mechanization was no idle threat. Even as the UFW battled growers on other fronts, 32,000 tomato pickers had been replaced by a mechanical tomato harvester. [23] UFW leaders also were aware that scientists at the University of California were experimenting with a mechanical lettuce harvester. [24] Experiences such as these indicated that if farmworkers were to succeed in their struggle, they would have to slow the pace of mechanization. Since the mid 1970s, the UFW has attempted to accomplish this goal, primarily by making a public issue of government support for further mechanization research.

While UFW efforts undoubtedly are the driving force behind the mechanization issue, its prominent place on today's political agenda also derives in part from a new willingness to question technological change in general. Recently, the public, academicians and government policymakers all have become more skeptical of technological fixes for society's problems. Not long ago we saw Congress reject the SST because of risks it posed to the environment. We also saw further research on Recombinate DNA temporarily halted while its ethical considerations were debated. And finally, we continue to see mass demonstrations against reliance on nuclear power as a solution to the energy crisis. Each of these events points to a new willingness to look at both the costs and benefits of new technologies before sanctioning their development or adoption. This more cautious view of technology may still be a minority opinion. And, the increasingly loud clamor for greater productivity ultimately may drown it out altogether. But, it nonetheless has made farmworker demands to assess the costs and benefits of further farm mechanization seem very logical.

Important as all of the above factors are in explaining the sudden prominence of the mechanization issue, they ultimately depend for their impact on a growing public awareness that technological choices are political decisions. Political economists long have argued that technological choices are political decisions primarily because they seriously affect the distribution of such important societal values as employment, income, health and safety. Such choices affect the distribution of these values because different technologies generate different costs and benefits for the various groups and classes in society. A decision to promote one kind of technology rather than another therefore often represents a decision to allocate benefits to certain groups or classes at the expense of others. The decision of American coal mining companies, for example, to adopt more capital intensive production technologies after World War II sharply reduced the demand for labor in mining. This, in turn, imposed extremely heavy costs on displaced miners. The new coal mining technology also imposed heavy costs on the communities which depended on miners' wages. Small businessmen in these communities often went bankrupt. Public employees also often lost their jobs. But, while the move to more capital intensive production technologies in the mines was imposing heavy costs on some sectors of society, it was generating important benefits for others. The most prominent beneficiaries of the new technology were firms which manufactured and sold it, larger mine operators who were able to use it to increase profits, and consumers who enjoyed the lower prices it allowed for. Coal miners who continued in the mines after the adoption of the new technology also benefited because it allowed coal companies to pay higher wages. [25]

Of course, to argue that technological choices are political decisions is to assume that mankind continues to exercise a significant degree of control over the process of technological change. And, this is an assumption that often has been challenged by students of technological change. These writers have argued that technology is now largely beyond any significant degree of human control. According to this view, technology provides its own momentum and responds only to the dictates of its own immutable logic. In the words of Jacques Ellul, a well-known proponent of this view:

> Technique pursues its own course more and more independently of man. This means that man participates less and less actively in technical creation, which, by the automatic combination of prior elements, becomes a kind of fate. Man is reduced to the level of catalyst. [26]

For Ellul and other proponents of the autonomous technology perspective, technological change is largely beyond societal control because it occurs primarily in accordance with some predetermined and unalterable script. Humans are not free to choose the technological innovations they desire. These innovations follow "logically" one after the other. Each new innovation is largely determined by those that have preceeded it. Because man is powerless to alter the sequence of these changes, his technological choices are of the most insignificant type. [27]

There is another sense in which the proponents of an "autonomous technology" do not see technological changes as social choices. It has to do with the "unintended consequences" which are inevitably generated by any new technology. According to this argument, there are always certain consequences of any innovation that cannot possibly be anticipated by society. This is because nature and society are so complex, and the interconnections between them are so involved that it is impossible to predict how a change induced in any part of this system will ultimately affect the other parts of the system. Who could ever have anticipated, for example, that the use of aerosol sprays might someday threaten the earth's protective atmosphere? If it is impossible to anticipate all of the consequences associated with any technological innovation, how, argue proponents of the autonomous technology school, can it be claimed that man exercises any significant control over the process of technological change? [28]

Critics of the "technological imperative" perspective are not completely sold on the above arguments. Certainly, man is never wholly free to choose between all possible alternatives. Technological choice is clearly limited by the boundaries of scientific knowledge. It is also clearly limited by the fact that man can never anticipate all the possible consequences that a particular technological change may precipitate. Allowing for these two constraints, however, man is still left to make important choices concerning the rate and kind of technological change he will encourage and accept in society. While scientific knowledge may set the parameters of technological choice, man still must choose the ends toward which he wishes to apply the discoveries of science. He also must decide which scientific advances to exploit. As various writers on this subject have already pointed out, "the existence of a technological capability does not inevitably lead to its exploitation." [29] Even the "unintended consequences" often associated with technological change still leave men significant latitude for social choice. Technological changes are rarely sharp and irreversibly shifts away from an old technology towards a new one. They are usually more continuous in nature, representing a gradual shifting away from one way of doing things toward another way of doing things. [30] Because these changes take place over many years, what originally might have been "unforeseen consequences" associated with a particular technological change become obvious. At this point, society must choose whether or not to accept these consequences. Apparently, a significant number of Americans now believe this choice is largely a political decision.

THE COSTS AND BENEFITS OF MECHANIZATION RESEARCH

The current controversy over further state support for mechanization research involves two different questions: (1) who controls the political decisions behind agricultural research, and (2) who benefits and loses from the mechanical innovations that come out of that research. This book addresses primarily the latter of these questions. It seeks to identify the major kinds of costs and benefits generated by the mechanization process and to analyze their distribution among different groups in society. In an attempt to analyze the distributive consequences of capital-intensive farming technologies, the book necessarily focuses at some length on the methodology presently being used to assess such impacts. It points out the major shortcomings in this methodology. The book then illustrates an alternate approach in a case study of the electronic tomato sorter, a new harvest technology currently being adopted in the processing tomato industry in California. Finally, the book draws upon some of its research findings to evaluate current public policy regarding technological change in agriculture.

By providing an answer to the question of who benefits and loses from agricultural mechanization, this book also will inevitably contribute to the debate over who controls the political decisions behind this change. If we find that these innovations have benefited agribusiness at the expense of other groups in society, we will have lent support to claims of an "unholy alliance" between agribusiness and the public sector research establishment. Alternately, if we find that these innovations clearly have benefited the wider public, we will have allayed fears about the sinister influence of corporate agriculture on public research priorities.

It is always somewhat risky, of course, to infer who controls the decision-making process behind a technological change from the distributive consequences it generates. In large part, this is because those who choose the new technology may never intend the consequences it generates. [31] Naturally, the greater the possibility of unintended consequences associated with any technological change, the less confident we feel about inferring anything regarding the nature of the decision-making process behind this change from its distributive consequences. Completely new and revolutionary changes in particular may have so many unintended consequences associated with them that it is impossible to make any reliable inferences about who controls the decisions behind them. Not all technological change is so revolutionary, however.

Rather than being first experiments with wholly untried technologies, many changes are simply innovative applications of already existing technologies. Sometimes these technologies have been in use for many years and have been studied extensively in prior applications. This kind of technological change is likely to generate relatively few unintended consequences. Because it does generate so few unintended consequences, we can make fairly reliable inferences about who controls the choices behind this change from the distributive consequences that it precipitates. Mechanization in agriculture is this type of technological change.

Footnotes

1. An indication of the issue's national appeal is suggested by the fact that within the past year it has made its way into local newspapers from Texas to Ohio, and has been the subject of major coverage in such national publications as The New York Times, Harpers, and Newsweek magazines.
2. The exact level of public support for mechanization research is not agreed upon by all parties to this debate. Opponents of this research fix the figure at well over $1 billion per year. Proponents fix the figure at less than half of that amount. For a discussion of conflicting claims over the dollar amounts involved, see the Lubbock Avalanche-Journal, May 25, 1979.
3. Jim Hightower, Hard Tomatoes, Hard Times (Cambridge, Mass.: Schenkman Publishing Company, 1973), p. 38.
4. These interpretations of the impact of mechanized agriculture are taken from numerous published and unpublished interviews with representatives of groups opposed to this process.
5. A representative statement of this perspective can be found in an editorial by J. B. Kendrick Jr., Director of the University of California's Agricultural Experiment Station, which appeared in California Agriculture, June 1977.
6. For a summary of the suit's charges, see the Berkeley Gazette, 17 January 1979.
7. Hightower, Hard Tomatoes, op. cit., p. 37.
8. An excellent summary of this and its competing interpretation of the decision-making process behind technological change can be found in Dale E. Hathaway, "Public Sector Research and Education and the Agribusiness Complex: Unholly Alliance or Socially Beneficial Partnership?" American Journal of Agricultural Economics, (August 1974), p. 993.
9. This perspective is known as the "Theory of Induced Innovation." It has been developed primarily by Vernon W. Ruttan and Yujiro Hayami. For a recent statement of this perspective, see Vernon W. Ruttan and others, "Factor Productivity and Growth: A Historical Interpretation," Induced Innovation, eds. Hans P. Binsqanger et. al. (Baltimore: John Hopkins Press, 1978), pp. 44-87.
10. Phillip Martin and Stanley S. Johnson, "Technology and Labor in Agriculture," unpublished paper presented at annual meeting of Western Economics Association, Anaheim, California, June 20-22, 1977, p. 1.
11. Wayne Rasmussen, "The Impact of Technological Change on American Agriculture, 1862-1962," Journal of Economic History, Vol. 22, 1962, pp. 578-599.

12. Ernesto Galarza, Farmworkers and Agri-business in California, 1947-1960 (Notre Dame: University of Notre Dame Press, 1977), pp. 66-69.
13. Ibid., p. 68.
14. M. Zahara and S. S. Johnson, "Status of Harvest Mechanization of Fruits, Nuts, and Vegetables," HortScience, Vol. 14(5), October 1979, pp. 578-582.
15. John A. Hopkins, Changing Technology and Employment in Agriculture, U.S. Department of Agriculture, May 1941, p. 60.
16. Technology on the Farm, A Special Report, U.S. Department of Agriculture, August 1940, pp. 10, 64.
17. Donald H. Grubbs, Cry From the Cotton: The Southern Tenant Farmers' Union and the New Deal (Chapel Hill, The University of North Carolina Press, 1971), pp. 1-135.
18. William H. Metzler, Farm Mechanization and Labor Stabilization, Part II, Technological Change and Farm Labor Use, Kern County, California, 1961, Giannini Foundation Research Report, No. 28, January 1965, p. 1.
19. Galarza, Farmworkers and Agri-business, op. cit., pp. 97-202.
20. Farm Labor Fact Book, U.S. Department of Labor, 1959, p. 155.
21. Ibid., p. 156.
22. Vernon Briggs Jr., The Mexico-United States Border: Public Policy and Chicano Economic Welfare, 1974, p. 9.
23. William Friedland and Amy Barton, Destalking the Wily Tomato, Research Monograph Number 15, Department of Applied Behavioral Science, University of California, Davis, June 1975, p. 41.
24. William Friedland, Amy Barton and Robert Thomas, Manufacturing Green Gold, Department of Applied Behavioral Sciences, University of California, Davis, July 1978, pp. 68-70.
25. Sar A. Levitan and Harold L. Sheppard, "Technological Change and the Community," Adjusting to Technological Change, eds., Gerald G. Somers et. al. (New York: Harper and Row, Publishers, 1963), pp. 169, 170.
26. The Technological Society, trans., John Wilkinson (New York: Alfred A. Knopf, 1964), p. 134.
27. Langdon Winner. Autonomous Technology (Cambridge: MIT Press, 1977), p. 70.
28. Ibid., pp. 89-97.
29. David and Ruth Elliott, The Control of Technology (London: Wykeham Publications, Ltd., 1976), p. 7.
30. Lambright, Governing Science (London: Oxford University Press, 1976), pp. 74-81.

31. The question of how much one can infer about who controls a decision-making process from the outcomes of decisions made therein continues to be a source of contention between "behavioralists" and "non-behavioralists." The former reject the validity of these inferences, claiming that the only way to differentiate actual power from potential power is to see it exercised in the decision-making process. Non-behavioralists, on the other hand, believe that power is not something that can always be seen. Therefore, they are more willing to infer the identity of the powerful from the outcomes of the decision-making process.

2
The Mechanization of Agriculture: An Overview of Its Costs and Benefits

What are the costs and benefits generated by the multitude of mechanical innovations adopted by American farmers? Which groups or social classes gain or lose the most from these innovations? In an effort to answer these questions and, by inference, to shed some light on the nature of the decision-making process behind these innovations, I have reviewed existing studies of agricultural mechanization both in the United States and Third World nations. In each case, I have focused on the costs and benefits generated by these innovations. Because this literature is extensive, I have not attempted to discuss here all the works appearing in it. Instead, I have tried to distill from representative studies in this literature only major generalizations and necessary qualifications regarding the costs and benefits generated by the principal kinds of mechanical innovations for the main groups in society affected by them. I also have attempted to point out how these distributive consequences may vary from one kind of change to another and from one commodity or social setting to another.

MECHANIZATION AND THE FARMER

The mechanization of farming has generated dramatic increases in agricultural productivity. It has allowed farmers to increase their output sharply while maintaining a constant or reduced level of labor and total inputs. [1] Logically, we might expect these sharp gains in production efficiency to generate significant increases in farmers' profits. Yet, most analysts believe that technologically generated increases in production efficiency have been largely translated into lower prices for consumers rather than larger profits for most farmers. [2] This point of view was outlined by Willard Cochran in 1958 when he emphasized that the continuing low incomes received by farmers of that period were due largely to low market prices resulting from the fact that technologically generated growth in aggregate farm supply had been outpacing aggregate demand. [3] While Cochran and most other economists believe that farmers as a whole have not benefited significantly from technological change, they have been careful to point out that farmers who adopt new agricultural technologies early generally enjoy handsome profits, at least in the short run. They enjoy handsome profits because they are able to use the new technologies to reduce their per unit costs and expand their output in the face of constant prices. In the long run, however, other farmers also desirous of these profits adopt the new technologies. The additional output of these latter adopters ultimately expands aggregate supply to a point where market prices decline, eliminating the opportunity for above average profits. Thus, while early adopters make attractive short-run profits by adopting new farm technologies, the average farmer is thought to be on a "technological treadmill" because by the time he adopts, market prices have already fallen to a point where his profit is no greater than it was prior to the advent of the new technology. [4]

Generally, large farmers are the first to adopt new technologies and capture the handsome short-run profits they generate. [5] Before farmers can adopt a new technology, they often must have access to large amounts of capital or other potentially costly prerequisites. Because larger farmers normally have greater access to capital and other technological prerequisities, they are more likely to adopt new technologies before smaller farmers. This is particularly true with regard to large-scale mechanical innovations which require large initial cash outlays. The newest mechanical tomato harvesters, for example, cost $140,000. [6] This amount is well beyond the means of most small farmers. Large-scale mechanical innovations are beyond the means of most smaller growers not only because of their high total costs, but also because they generally have important economies of scale associated with their use. This means that they are more costly per unit of output for smaller farmers than for larger farmers. The higher per unit of output costs associated with large-scale mechanical innovations often have forced smaller growers to abandon entirely production of commodities affected by such technologies. Adoption of the mechanical cotton harvester, for example, forced roughly 54 percent of the smaller growers out of that industry. [7] Similarly, adoption of the mechanical tomato harvester in California reduced grower ranks in this industry by approximately 85 percent. Most of those leaving were smaller growers having acreages far below the capacity of this new technology. [8]

Elimination of the smaller grower through the process of technological advance provides an important long-term benefit to larger growers adopting these innovations. It often facilitates the organization of remaining larger growers into effective grower organizations capable of exercising significant market power or governmental influence. Adoption of the cotton harvester, for example, is thought to have contributed greatly to organizing remaining larger growers into a producer organization capable of lobbying effectively for cotton price supports. [9] Similarly, adoption of the mechanical tomato harvester is thought to have greatly facilitated organization of remaining growers for purposes of collective price bargaining with major food processors. [10] In each of these cases, elimination of numerous smaller growers through technological change facilitated the organization of remaining larger growers. These organizational advantages were then translated into concrete monetary gains for those who remained in the industry in the machine era.

While large farmers are better prepared to capture the benefits of mechanization than their smaller counterparts, both groups of producers must face certain unavoidable costs that accompany the move to a more mechanized style of farming. The most important of these is an increased vulnerability to the vissicitudes of the market place. [11] Mechanization inevitably leads farmers to become more specialized and more commercial than they would be with a more labor-intensive style of farming. [12] Because the cost of new large-scale farm machinery is so high, farmers must use it as much as possible to recoup their fixed costs. If, for example, a farmer invests $140,000 in a new tomato harvester, he must use this new technology for several seasons if he is to recover the cost of his initial investment. He can no longer afford to grow tomatoes one year and shift to an alternate crop the next year if the latter appears slightly more profitable. Instead, he must continue growing tomatoes for the entire useful life of his investment, even if the price of tomatoes falls off sharply. [13] This increased specialization and reduced flexibility that accompanies mechanization necessarily increases the vulnerability of the farmer to unpredictable market forces. For the same reason that mechanization leads to an increase in farm specialization, it also leads to an increase in farm commercialization. Once a farmer invests in a costly piece of harvesting equipment, he cannot afford to grow anything less than the harvesting capacity of the machinery involved. To do so would force him to forego the economies of scale inherit in the new technology. As a result, he grows more of a crop than he might choose had he not been so fully mechanized. An indication of this increasingly commercialized character of mechanized agriculture is found in the fact that there has been a 40 percent decline in the proportion of farms marketing less than $2,500 worth of produce (in constant dollars) since 1960. [14] Of course, the more commercialized farmers become, the more they are threatened by sharp changes in the market prices of their products. [15]

Mechanization has done more than increase a farmer's vulnerability to the market price of his output. It also has made him more dependent on those who supply him inputs. The cost of new farm machinery has been increasing more rapidly than the farmer's ability to pay for it. As a result, he has had to borrow heavily from banks and large farm machinery companies to maintain his place on the technological treadmill. An indication of this borrowing trend is suggested by the fact that in 1950 U.S. farmers could have paid off their total debt with 75 percent of their annual income. Today, they would need over three year's income to pay off their total debt. The more farmers borrow, of course, the more independence they give up to their creditors. Some banks in the Midwest, for example, have forced farmers to continue a chemical rather than an organic approach to farming even when the farmer himself prefers the latter. [16]

To this point, I have focused on the role of technological change mainly as it affects farmer income by influencing prices they receive in the commodity markets. Professors Herdt and Cochrane have reminded us, however, that farmers also may gain or lose from technological change in their role as land owners. In 1966, they argued that under conditions of government supported farm prices, retiring farmers and landowners who sell farmland have been the primary beneficiaries of technological change in agriculture. Under these conditions, technological changes lower farmers' production costs and increase their profits per acre. This encourages farmers to expand their acreage in order to increase their total profits. In pursuit of these increased profits, farmers bid up the price of the land, distributing the gains of technological change to those selling farmland. These landowner gains may even eliminate the consumer gains thought to be generated by technological change under more competitive market conditions. This occurs if rising land costs push up production costs to a point where the costs saved by the new technology are completely cancelled out. Price supports, of course, are crucial to this interpretation of winners and losers. When there are no price supports, lower per unit cost of production leads to increases in supply which, in turn, brings lower commodity prices. The lower commodity prices make land less attractive, reducing and perhaps even eliminating entirely the gains to landowners resulting from technological change. [17]

MECHANIZATION AND FARM LABOR

Mechanization in agriculture affects farm laborers in two very direct ways. It drastically reduces their employment opportunities, and it significantly increases the productivity of those workers who continue to find employment on more mechanized farms. Due to increased mechanization, almost all major farm commodities can be produced today with only a fraction of the farm labor they required fifty years ago. In 1930, production of 100 bushels of wheat required 70 hours of farm labor; today it requires only 8 hours. In 1930, production of 100 bushels of corn required 127 hours of farm labor; today it requires only 6 hours. In 1930, production of a bale of cotton required 252 hours of farm labor; today it requires 23 hours. [18] Similarly impressive reductions in farm labor requirements for most commodities have drastically reduced the total demand for farm labor in the U.S. In the 1910-1914 period, the average annual number of man hours of labor used for farm work in the United States was 23 billion. Today this figure is less than 6 billion. [19] The drastic cut back in the total demand for farm labor indicated by the above figures has taken place in spite of the fact that total farm output in the United States has increased 63 percent during the past thirty years. [20]

The adoption of major mechanical innovations in agriculture inevitably has brought about a massive displacement of farm labor. But there is no simple inverse relationship between mechanization and reduced demand for farm labor. Certain mechanical innovations increase total output of the commodity they affect as well as substitute capital for labor. Adoption of the tractor, for example, increased total farm output in the United States by releasing land formerly needed to maintain draft animals for production of human foods. This increased cropland raised total farm output and tended to increase the demand for farm labor. But, because adoption of the tractor also increased the number of acres that a single farmworker could cultivate, it simultaneously reduced the per acre demand for farm labor. The net effect of the tractor on total demand for farm labor, therefore, depended on the size of its output effect relative to it's labor substitution effect. [21] What was true for the tractor also is true for other mechanical innovations in agriculture. [22] In each case, their net effect on farm labor is a function of the tradeoffs between their output effect and their labor substitution effect. Certain kinds of farm machinery, of course, are likely to have more significant output effects than others. Farm machines designed to permit more extensive cultivation, for example, are likely to have more important output effects than large-scale harvesting equipment. Even farm machinery having significant output effects, however, has consistently precipitated displacements in farm labor. This is because the labor substitution effects associated with these innovations consistently have dwarfed their output effects. In spite of the tractor's sizable output effect, for example, its adoption in the plantation South displaced approximately 500,000 farm laborers. [23]

Mechanization has seriously eroded employment for farm laborers as a whole. But, it has hit some elements in this work force harder than others. In general, mechanization has displaced non-white farm labor at a faster rate than its white counterpart. Mechanization of the rural South in the 1940s, for example, displaced black sharecroppers, tenants and farm laborers at a much faster rate than comparable groups of whites. [24] Also, mechanization has brought much greater dislocation to seasonal farmworkers than to full-time, hired workers. [25] Between 1959 and 1973, there has been a 40 percent drop in the number of hired laborers employed on American farms. But, the number of full-time workers in this hired labor group has actually increased slightly. [26] This means that the full brunt of technological displacement has been borne by seasonal workers.

Technological displacement doesn't necessarily impose heavy costs on workers unfortunate enough to experience it. According to Edwin Mansfield, a noted student of technological change, this is because "there is a growing feeling that society, which is by and large a beneficiary of technological change, has an obligation to minimize the losses and assist the readjustment of those who are hurt." [27] Unfortunately, this "feeling" of which Mansfield wrote in 1968 has not yet evidenced itself very strongly with regards to the victims of agricultural mechanization. The most important social programs available to cushion unemployed workers are the State Unemployment Insurance programs. Yet, as late as 1975, only three states in the United States provided unemployment insurance for farmworkers. [28] Since that date, only the states of California, Minnesota, and Hawaii have changed their policies to make unemployment insurance available to all farmworkers. Government slowness to provide even the most basic adjustment allowance largely explains why mechanization has consistently imposed extremely heavy economic costs on American farmworkers.

While mechanization has radically reduced the total demand for farm labor, it has sharply increased the productivity of the farmworkers who remained employed in agriculture. The clearest single indication of this increased farmworker productivity is provided by the fact that total farm output in the United States has increased 63 percent during the last thirty years in spite of a 70 percent reduction in manhours worked by farm labor. [29] How much of these dramatic increases in farm labor productivity have been translated into higher wages for retained farmworkers is unclear. The majority of mainstream economists have long presumed that wages are directly tied to worker productivity. This presumption leads them to suggest that increases in workers' productivity will necessarily lead to increases in workers' wages. [30] An increasing number of political economists, however, have begun to challenge this hypothesis. Their perspective on this question is perhaps best represented by the remarks of Ernest Mandel who has argued as follows:

> The rise in real wages does not follow automatically from the rise in the productivity of labour. The latter only creates the possibility of such a rise...For this potential increase to become actual, two interlinked conditions are needed: a favorable evolution of relations of strength in the labor market...and effective organization...of wage workers which enables them to abolish competition among themselves and so to take advantage of these favorable market conditions. [31]

Empirical support for this position has been provided, in part, by Barry Bluestone's study of the characteristics of low-wage industries. According to Bluestone:

> The productivity gains in low-wage industry are not reflected in the relative wage rate changes in low-wage industry. Rather than contributing to higher wages, productivity increases are either being absorbed into broader profit margins or otherwise into lower prices due to raging competition. [32]

Although Bluestone's observations regarding productivity gains and wage increases were based primarily on manufacturing industries, they help to explain the relationship between these two variables in agriculture as well. Between 1950 and 1970, the productivity of farm labor increased by 122 percent. [33] But, during the same period farm wages increased by only 84 percent. [34] These figures indicate that although an important part of the gains generated by increased farmworker productivity have gone to farm labor in the form of higher wages, a significant part also has been translated into broader farmer profit margins or passed on to consumers in the form of lower prices. These figures also suggest that mechanization has had a dual effect on the wages received by farmworkers retained on more mechanized farming operations. On the one hand, it has created the conditions for a major increase in their wages. On the other hand, it has been a major force in preventing them from organizing to capture this increase. This latter effect stems from the critical role mechanization has played in undermining union organizing attempts by farmworkers. According to Michael Perelman, for example:

> Just when workers began to organize successfully in the California tomato fields, scientists and engineers at the Davis Campus of the University of California coordinated their energies to put the finishing touches on a mechanical tomato harvester. [35]

This view of mechanization as the ultimate weapon against union organizing attempts has since been confirmed in a series of interviews with tomato growers themselves. [36]

TECHNOLOGICAL CHANGE AND THE CONSUMER

Until quite recently most students of technological change in agriculture agreed that consumers had been the primary beneficiaries of farm mechanization. They pointed out that by sharply increasing agricultural productivity, mechanization helped to expand the total food supply in the face of relatively inelastic consumer demand. This, in turn, brought major benefits to consumers in the form of increased consumption and lower prices. [37] While many analysts still subscribe to this interpretation of mechanization's impact on the consumer, they are increasingly careful to acknowledge that its accuracy rests heavily on the assumption that the prices of commodities affected by these innovations are determined largely by the forces of supply and demand operating in competitive markets. Many observers of American agriculture still consider this a realistic assumption in general. [38] They acknowledge, however, that this may not be the case for all farm commodities. [39] In these cases, they have cautioned that consumer gains from technological change may be seriously eroded or completely usurped by other groups in the food chain. [40]

Although many analysts continue to believe that in general consumers are the primary beneficiaries of agricultural mechanization, a growing number of economists and rural sociologists have challenged this view. At the heart of their argument is the belief that the impersonal forces of supply and demand operating in competitive markets are no longer the sole determinants of farm commodity prices. They emphasize the increasing degree of economic concentration occurring at all levels of the food industry. Grower organizations, labor unions and giant corporations at the producer, processor and retail levels are acquiring sufficient market power to influence prices. Commenting on the existence of economic concentrations at the producer level, Nick Kotz has noted, for example, that:

> Twenty large corporations now control poultry production (in the U.S.)...United Brands, Purex, and Bud Antle, a company partly owned by Dow Chemical, dominate California lettuce production. [41]

Other analysts have made similar observations about the existence of market power at the processor and retail levels. A University of California economist, for example, recently told a State Senate hearing that Safeway Stores, Inc. and other giant supermarket chains use their enormous market power to cut the prices that cattlemen receive for their beef by 10 percent. At the same time, he claimed, they are able to inflate the prices consumers pay for this beef in the supermarket by another 10 percent. [42] These and numerous observations of the exercise of market power by organizations at every level of the food industry have led many analysts to reject the assumption that agricultural prices are determined competitively and, with it, the conclusion that the benefits of technological change inevitably accrue to consumers. Instead, they have suggested that the distribution of benefits generated by new agricultural technologies needs to be viewed more in terms of social and economic conflict between the major groups within the food production and distribution sectors. According to this perspective, the ultimate distribution of benefits generated by each type of technological change "will depend on how much of the economic surplus can be extracted by each group." [43]

Other analysts also have pointed out that consumer gains from technological changes may be seriously eroded if the prices of commodities affected by these changes are determined by government support programs or in the international market. [44] Under both of these conditions, increases in the aggregate supply of commodities affected by technological changes may not lower domestic prices or increase domestic consumption. In these cases, the benefits generated by technological change will accrue primarily to producers rather than consumers. Support for this conclusion was provided in part by an analysis of the welfare effects of new cotton varieties adopted in Brazil. In this analysis, Ayer and Schuh observed that the majority of benefits resulting from the improved varieties were captured by cotton producers in Sao Paulo rather than passed on to domestic consumers in the form of lower prices because an important part of that cotton was sold on the international market, leaving domestic supply and price relatively unaffected. [45] While Ayer and Schuh's observation regarding the distribution of technological benefits was based on a biological innovation, it applies to mechanical innovations as well.

There are two other factors, in addition to those I have already mentioned, which may affect the amount of benefits that consumers can expect from technological changes in agriculture. Several studies have demonstrated that, other things being equal, the less sensitive consumer demand is to price changes of a commodity, the greater the consumer benefits that may be generated by an innovation affecting that commodity. [46] When a technological innovation occurs in a commodity industry, it stimulates an increase in the supply of that commodity on the market. This expanded supply of the commodity will bring about a decline in its market price. The extent to which the commodity's price declines depends primarily on how much of the new supply is taken up by additional purchases of the commodity at lower prices. The smaller the rate of increase in quantity demanded relative to the rate of price decline (the smaller the price elasticity of demand for the commodity), the lower the new equilibrium price will be and greater will be the benefits consumers derive from the new technology. Since the elasticity of demand varies between agricultural commodities, the amount of consumer benefits generated by innovations affecting them also will differ. In general, technological advances in the production of subsistence crops are thought to generate larger potential benefits for consumers because they evidence less elasticity of demand than many of the specialty commodities. [47]

A final consideration affecting the amount of consumer benefits generated by new farm machinery arises from the fact that different farm commodities are consumed by different segments of consumers. Commodities such as potatoes or onions are staples consumed in roughly equal amounts by consumers in all major income classes. Other commodities, such as cream or certain fresh fruits and vegetables, are largely luxury items disproportionately consumed by a much smaller number of more economically privileged shoppers. Finally, a few commodities, such as cabbage or dried beans, may be disproportionately consumed by poorer segments of the buying public. [48] Technological advances affecting those commodities that are most widely consumed are, of course, likely to generate larger consumer benefits than are innovations that affect commodities consumed largely by narrow segments of the public. [49]

Regardless of the value of the economic benefits which consumers ultimately reap from mechanical innovations in agriculture, their gains must be discounted by two important costs. The first of these costs is the constraint mechanical innovations place on consumer choice. This narrowing of consumer choice stems from the fact that these innovations encourage farmers to produce more uniform crop varieties. [50] A farmer substituting farm machines for farm labor must chose between only those relatively few plant varieties specifically developed for machine handling. [51] This results in a relatively uniform crop that limits the range of consumer choice. According to many consumers, mechanization also yields a crop which is relatively "tasteless." [52] They complain increasingly of "rubber tomatoes" or "oranges that don't taste like oranges." They argue that in an effort to make crops more resistant to machine handling, farmers have increasingly sacrificed "tasty" produce for more thick skinned "hardy" produce.

MECHANIZATION AND THE LINKAGE INDUSTRIES

Farmers, farmworkers and consumers are the major groups directly affected by mechanization in agriculture. But, this process also affects many other groups in society whose fortunes are directly linked to it. Most obvious of these other groups are owners and employees of companies that manufacture and finance farm machinery. While it seems likely that actors in both of these linkage activities have been major beneficiaries of farm mechanization, very little analysis of these presumed benefits appear in the literature. One exception to this general trend is an unpublished study that I made to ascertain how much new employment was created in the farm machinery industry by introduction of the mechanical tomato harvester in California in the mid 1960s. Relying on information supplied by the principal manufacturers of the harvester, I estimated that production of this innovation created 414,000 hours of labor annually at an average hourly wage of $5.50. The total value of this "labor benefit" amounted to $2,277,000 or 4.8 percent of the "labor cost" imposed on domestic field workers displaced by the harvester. [53]

Those who build and finance new farm machinery are not the only actors indirectly affected by mechanization in agriculture. This process also affects the fortunes of those in the food processing and transportation industries. And, as I noted earlier, many analysts have argued that the superior market power exercised by the largest food processing firms in particular enable them to extract the lion's share of economic surplus created by technological advances on the farm. This view leads us to believe that agricultural mechanization has generated some combination of increased wages and profits in the food processing industry. But, no analysis exists in the literature linking profit or wage increases in this industry to technological advances on the farm.

While the literature does not contain any studies specifically tying increased wages and profits in the food processing industry to mechanization, several analysts have linked new employment opportunities in the industry to increased mechanization at the farm level. In a study of the processing tomato industry, for example, Jon Brandt estimated that the mechanical tomato harvester created 110,000 additional weeks of work in the food processing industry, and 5,714 additional weeks of work in the farm transportation industry to compensate for the 325,000 weeks of work it eliminated in the tomato fields. Brandt's estimates of the amount of additional employment created by the tomato harvester were based on the claim that this new technology allowed growers to plant an additional 51,000 acres of tomatoes in areas of California thought to be "labor shortage areas." [54] This claimed existence of farm labor shortage areas in California is, of course, very debatable and, no doubt, would be seriously contested by farm labor groups in the state.

MECHANIZATION AND SOCIETY

Mechanization in agriculture has generated some very direct costs and benefits to different groups in society. But, it also has generated several more diffuse and indirect costs and benefits for society as a whole. Perhaps the most important of these more diffuse societal benefits is the contribution agricultural mechanization has made to the nation's economic development. It is widely agreed that by releasing massive numbers of laborers from the farm, mechanization simultaneously reduced the prices Americans had to pay for food and provided the growing manufacturing sector with a needed reservoir of workers. Because the jobs provided in industry typically were better paying than those provided by agriculture, farm mechanization raised the economic standard of living for both those who remained on the farm and those who left it. [55]

While there is considerable agreement on the historical contribution of agricultural mechanization to the nation's economic development, there is some disagreement over its contribution to economic development in the current era. This disagreement stems largely from a debate over the impact of mechanization in general on employment. Optimists in this debate generally have acknowledged that mechanization creates short-term unemployment problems for different groups in society affected by it. But, they have emphasized that mechanization creates much more employment than it abolishes. According to proponents of this point of view, mechanization creates employment in at least three major ways: (1) By lowering costs and prices in one industry, it allows consumers to buy either more of the goods concerned or more goods and services in other industries. This, in turn, means more employment. (2) Mechanization creates a demand for machines which, in turn, creates a need for workers who build, design, install and maintain them; and (3) Mechanization creates the demand for new products and services which, in turn, creates more employment. Invention of the automobile, for example, created a tremendous demand for a wide range of new products (tires, machine parts, radiators, etc.) that were not formerly needed by society. Believing that mechanization invariably creates more employment than it abolishes, and exerts a downward pressure on prices, optimists feel that mechanization in any era must contribute positively to economic development. [56]

Not all who have studied this issue are so sanguine about the impact of mechanization on employment and national economic development. Acknowledging mechanization's employment-creating potential, skeptics, nonetheless, have questioned whether this process, operating in a free market setting, necessarily creates more employment than it eliminates. Some have argued that without rapid expansion of the public service sector and defense related production, which has characterized the U.S. since the close of World War II, mechanization would have resulted in unemployment rates significantly higher than those actually seen. [57] Others have linked the ability of the nation's economy to absorb technologically displaced workers to aggregate growth rates and suggested that as the latter have slowed, mechanization efforts have had an increasingly detrimental effect on society. [58] In effect, skeptics have argued the mechanization in general is not intrinsically beneficial or harmful to a nation's economic development. Its ultimate impact depends on such intervening variables as existing unemployment rates, government efforts to facilitate movement of dislocated laborers from one industry or area to another, and the extent of competition in the industries affected.

There is little hope for a definitive resolution to the debate over mechanization's current contribution to society's economic development. There is, however, little doubt that any economic benefits this process generates must be balanced against the continuing loss of economic and, perhaps, political independence it imposes on most members of society. Mechanization, in nearly all forms, has consistently driven smaller businessmen out of the industries affected by it. It also has made it increasingly difficult for all but the more privileged members of society to gain entry into many fields of economic activity. In short, it has been a major force in transforming a nation of small independent businessmen and farmers into a nation of employees working in another man's shop. [59] That this has sharply curtailed the capacity of most men to shape their work environment can scarcely be doubted. But, to the extent that making one's living by taking directions from others erodes the self-confidence that one develops making one's own decisions, mechanization may also have worked against maintenance of the democratic character.

CONCLUSION

I began this chapter searching for major generalizations regarding the cost and benefits generated by mechanization in agriculture for the main groups in society affected by it. But, the only certain generalization which has emerged from the multitude of studies on this issue is that there are no certain generalizations concerning the distribution of these costs and benefits. Instead, we have seen that the distribution of gains and losses varies a great deal from one technology to another, from one commodity industry to another, and from one socio-economic setting to another. What degree of competition exists in the various markets linking farmers to consumers? How well organized is labor at the various levels of the food chain? And, how equally are capital and other technological prerequisites distributed among potential adopters of the new technology? The answers to these questions may vary tremendously from one innovation to another. Because the answers to these questions may vary greatly, so also may the costs and benefits of mechanization in agriculture.

Footnotes

1. Harold O. Carter and Warren E. Johnson, "Agricultural Productivity and Technical Change: Some Concepts, Measures and Implications," Technological Change, Farm Mechanization and Agricultural Employment, Division of Agricultural Sciences, University of California, July 1978, p. 74.
2. This generalization is supported by a wide range of economists. See, for example, Theodore W. Schultz, "A Policy to Redistribute Losses From Economic Progress," Journal of Farm Economics, 59 (May 1977), p. 245.
3. Willard Cochrane, Farm Prices: Myth and Reality (Minneapolis: University of Minnesota Press, 1958), p. 85.
4. Ibid., pp. 85-93.
5. Jurg Bieri, Alain De Janvry and Andrew Schmitz, "Agricultural Technology and the Distribution of Welfare Gains," American Journal of Agricultural Economics, 54 (December 1972), p. 807.
6. This figure is provided by the president of the Black Welder Farm Machinery Corporation, a major manufacturer of tomato harvesters in California.
7. Seymour Melman, "An Industrial Revolution in the Cotton South," The Economic History Review, 2 (1949), p. 67.
8. William Friedland and Amy Barton, Destalking the Wily Tomato, Research Monograph Number 15, Department of Applied Behavior Sciences, University of California, Davis, June 1975, p. 54.
9. Seymour Melman, "Industrial Revolution in Cotton," op. cit., pp. 68-70.
10. Friedland and Barton, Destalking the Wily Tomato, op. cit., pp. 56-59.
11. Michael Perelman, Farming for Profit in a Hungry World: Capital and the Crisis in Agriculture (New York: Universe Books, 1977), p. 85.
12. Ibid., p. 89.
13. This point is elaborated on by Wen S. Chern and Richard E. Just in Econometric Analysis of Supply Response and Demand for Processing Tomatoes in California, Giannini Foundation Monograph Number 37, University of California, Division of Agricultural Sciences, September 1978, p. 86-66.
14. Alvin L. Bertrand, "Rural Social Organizational Implications of Technology and Industry," Rural U.S.A. Persistence and Change, ed., Thomas R. Ford (Ames: Iowa State University Press, 1978), p. 78.
15. Luther G. Tweetern, "The Income Structure of Farms by Economic Class," Journal of Farm Economics, 47 (May 1965).
16. Perelman, Farming for Profit, op. cit., pp. 89-90.

17. Robert W. Herdt and Willard Cochrane, "Farm Land Prices and Farm Technological Advance," Journal of Farm Economics, 48 (May 1966), p. 262.
18. U.S.D.A., Agricultural Statistics, 1976.
19. Tolley, G. S. and B. M. Farmer, "Farm Labor Adjustments to Changing Technology," in Farm Labor in the United States, ed. C. E. Bishop (New York: Columbia University Press, 1967), p. 43.
20. Carter and Johnson, "Agricultural Productivity," op. cit., p. 24.
21. Technology on the Farm, U.S. Department of Agriculture, 1940, p. 65.
22. For a discussion of the output effect of the mechanical tomato harvester, see Jon Brandt, An Economic Analysis of the Processing Tomato Industry (Ph.D. dissertation, University of California, Davis, 1977), pp. 55-67.
23. Technology on the Farm, op. cit., p. 64.
24. Ford, Arthur M., Political Economics of Rural Poverty in the South (Cambridge, Mass.: Ballinger, 1973), pp. 20, 21.
25. Goss, Kevin F. and Richard O. Rodefeld, "Consequences of Mechanization in U.S. Agriculture," unpublished paper presented at the Rural Sociological Society Annual Meetings, Madison, Wisconsin, September 1977, p. 10.
26. Bertrand, op. cit., "Implications of Technology and Industry," p. 79.
27. Edwin Mansfield, The Economics of Technological Change (New York: W. W. Norton and Company, Inc., 1968), p. 135.
28. The Los Angeles Times, 9 September 1976.
29. Carter and Johnson, "Agricultural Productivity," op. cit., p. 97.
30. See, for example, Melvin Reder, "Wage Structure Theory and Measurement," cited in David M. Gordon, Theories of Poverty and Underemployment (Lexington, Mass.: D. C. Heath and Company, 1972), p. 29.
31. Ernest Mandel, Marxist Economic Theory (New York: Monthly Review Press, 1968), p. 145.
32. Barry Bluestone, "The Characteristics of Marginal Industries," in David M. Gordon, Problems in Political Economy: An Urban Perspective (Lexington, Mass.: D. C. Heath and Company, 1971), p. 103.
33. E. Philip LeVeen and Jerry Moles, California Agriculture in Transition, Monograph Number 1, California Agricultural Policy Seminar, University of California, Davis, 1979, p. 5.
34. Ibid.
35. Farming for Profit, op. cit., pp. 73-74.

36. Scheuring, Ann F., "From Lug Boxes to Electronics: A Study of California Tomato Growers and Sorting Crews," Monograph Number 3, California Agricultural Policy Seminar, University of California, Davis, 1978, p. 19.
37. This generalization is supported by a wide range of economists. See, for example, Cochrane, op. cit., p. 85, or Carter and Johnson, op. cit., p. 74.
38. See, for example, Daniel Suits, "Agriculture," in The Structure of American Industry, 5th ed., Walter Adams, ed. (New York: Macmillan Publishing Company, Inc., 1977), p. 2.
39. For an overview of economic concentration levels in principal farm commodity industries, see Staff Report on Agricultural Cooperatives, Federal Trade Commission, Bureau of Competition, September 1975, pp. 108-118.
40. Bieri et. al., "The Distribution of Welfare Gains," op. cit., p. 808.
41. Nick Kotz, "Agribusiness," Radical Agriculture, ed. Richard Merrill (New York: New York University Press, 1976), p. 43.
42. The Sacramento Bee, 28 July 1977.
43. Bieri et. al., "The Distribution of Welfare Gains," op. cit., p. 808.
44. Yujiro Hayami and Robert Herdt, "Market Price Effects of Technological Change on Income Distribution in Subsistence Agriculture," American Journal of Agricultural Economics, 59 (May 1977), p. 246.
45. Harry Ayer and G. Edward Schuh, "Social Rates of Return and Other Aspects of Agricultural Research in Sao Paulo, Brazil," American Journal of Agricultural Economics, 54 (November 1972), pp. 559-566.
46. Reed Hertford and Andrew Schmitz, "Measuring Economic Returns to Agricultural Research," in Resource Allocation and Productivity in National and International Agricultural Research, eds. Thomas M. Arndt, Dana G. Dalrymple and Vernon W. Ruttan (Minneapolis: University of Minnesota Press, 1977), pp. 155-156.
47. Ibid.
48. Suits, "Agriculture," op. cit., p. 10.
49. For a discussion and practical example of the importance of this factor, see Grant Scobie and Rafael Posada T., "The Impact of Technical Change on Income Distribution: The Case of Rice in Colombia," American Journal of Agricultural Economics, 60 (February 1978), pp. 58-92.
50. This observation was made long ago by S. Giedion in his Mechanization Takes Command (New York: Oxford University Press, 1948), pp. 132-134.
51. Ibid

52. Charles Handy and Martin Pfaff, "Consumer Satisfaction with Food Products and Marketing Services," U.S.D.A., Economic Research Service, Agricultural Economic Report No. 281 (March 1975), pp. 15-17.
53. Barry L. Price, "A Review of the Consequences Associated with Adoption of the Mechanical Tomato Harvester," unpublished paper written June 1978, see pp. 21-25.
54. Brandt, "An Economic Analysis," op. cit., pp. 55-67.
55. Rasmussen, "The Impact of Technological Change," op. cit., p. 590.
56. Canadian Labor Congress, Probable Effects of Increased Mechanization in Industry, Royal Commission on Canada's Economic Prospects, September 1956, pp. 28-33.
57. Frederick Pollock, Automation: A Study of Its Economic and Social Consequences, Trans. W. O. Henderson and W. H. Chaloner (New York: Praeger, 1957), pp. 54-58.
58. Goss and Rodefeld, "Consequences of Mechanization in U.S. Agriculture," op. cit., p. 7.
59. Perelman, Farming for Profit, op. cit., pp. 70-71, 80-81, 89.

3
The Method Behind the Madness of Current Mechanization Policy

The literature review on technological change in agriculture conducted in chapter 2 indicated that the costs and benefits generated by this process will vary considerably according to: (1) the particular technology involved, (2) the specific crop affected, (3) the socio-economic setting in which the change occurs, and (4) the response of the State to this change. On the basis of this observation, one might expect analysts of this process to recommend to government policymakers the need for a piece-meal approach in dealing with these changes. By this, I mean that one might expect them to recommend that government support for and approval of new agricultural technologies be contingent upon an assessment of the costs and benefits that are likely to be generated by each of these innovations. But, the vast majority of agricultural scientists and economists, the primary policy advisors on this issue, have vociferously rejected this piece-meal approach to technological change in agriculture. Instead, they have recommended strongly that government continue its heretofore unquestioned support for all forms of farm innovations. They have rationalized this policy recommendation largely by reference to a handful of "Social Rate of Return" studies done on specific agricultural innovations adopted in both the United States and various Third World nations. [1]

The principal argument of this chapter is that existing Social Rate of Return studies on new farm technologies have failed completely to portray the full impact of these technological changes on American society and, therefore, have supported policy prescriptions which are no longer in the public interest. By ignoring completely some of the most telling social costs generated by these innovations, existing cost-benefit analyses have consistently overestimated the social benefits associated with these changes. This is because they have: (1) consistently emphasized the total magnitude of the costs and benefits involved rather than the distribution of these gains and losses; (2) routinely weighted equally all the gains and losses generated in this process, rather than according greater weight to the losses suffered by those at the lower end of the income scale; (3) continued to base their conclusions on overly optimistic assumptions regarding the existence of largely competitive markets, rather than allowing for the increasing likelihood of limited competition in markets throughout the food chain; and (4) ignored the impacts these changes might have had on employment, wages and profits in both linkage and competing industries. Finally, because analysts conducting these studies have focused almost wholly on technological innovations which are widely viewed as "success stories," rather than examining a representative cross section of these innovations, their cumulative results suggest an even more distorted picture of technological change than do the individual studies when viewed separately. In short, then, the overly optimistic view of technological change in agriculture suggested by existing analyses of the costs and benefits of specific farming innovations can be traced largely to the methodology of the studies involved, and not to the consequences these changes actually precipitate. For this reason, it is important to look more closely at the methodology underlying these studies and to discuss more specifically the shortcomings of this approach.

At the heart of all Social Rate of Return (SRR) studies are two major steps. First, relying on records of time and expenses that agricultural scientists have devoted to developing a new technology, researchers calculate society's dollar investment in the innovation in question. Second, relying heavily on the concepts of producer and consumer surplus, researchers attempt to calculate the dollar value of annual net benefits the new technology generates. Juxtaposing this latter estimate against the dollar value of the research investment then allows researchers to calculate the "Social Rate of Return" associated with the technology under study.

Most critical to an understanding of this kind of cost-benefit analysis are the concepts of consumer and producer surplus used in step two of this approach. As used in these studies, consumer surplus refers to the amount of money that buyers save by purchasing a particular commodity at its market price rather than at the price they would have to pay for it under conditions of perfect price discrimination. [2] The concept is best illustrated by reference to a typical commodity demand curve like that pictured in figure 3.1. As is the case with any such demand curve, this curve represents the varying quantities of a particular commodity that buyers are willing to purchase as the commodity's price changes. It illustrates that even if the price of a commodity is quite high, P(4) for example, someone or few buyers can be expected to purchase a small quantity of it because they value the good that much. Similarly, if the price fell slightly

FIGURE 3.1

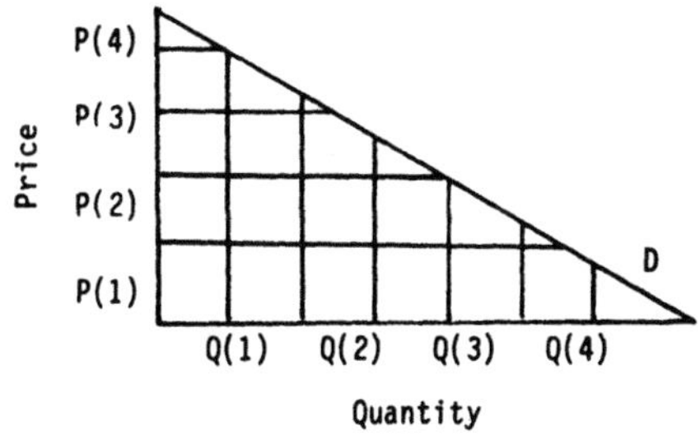

but remained relatively high still, P(3) for example, a slightly larger number of buyers can be expected to buy a slightly greater quantity of it because they place precisely that much value on it. Thus, in figure 3.1, those few buyers who placed a value of P(4) on the commodity would actually have to pay this amount to obtain it. Those who placed a value of P(3) on it would actually have to pay this amount to obtain it. And so on down the line to P(1). Of course, in an actual market situation each buyer, regardless of how much he actually values the good in question, pays the same price for it, and this price is determined by the interaction of the commodity's supply and demand. Referring again to figure 3.1, this means that those buyers who would have had to pay as much as P(4) or P(3) for this commodity under conditions of perfect price discriminatin can obtain it for as little as P(1) if available supply of and demand for the good have established this lower price as the market price. In such a case, those buyers who would have been willing to pay the higher prices for the good in question, but who were able to buy it at a lower price in the market setting, have enjoyed a monetary savings. The aggregation of each of these individual buyers' savings is defined as the consumer surplus. It is simply the aggregated difference between the price that each buyer would have been willing to pay for a commodity, under conditions of perfect price discrimination and that price that he actually paid for it in the market situation. [3] This sum equals the area of the lined triangle under the demand curve and above the price line in figure 3.1.

FIGURE 3.2

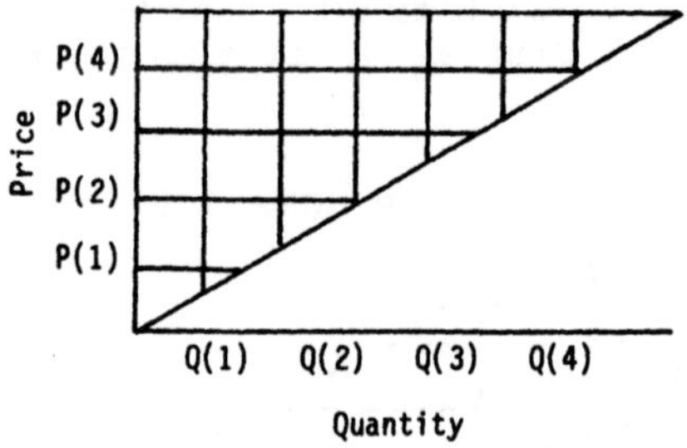

Producer's surplus is conceptually very similar to the notion of consumer surplus, but it applies to the savings that sellers receive from selling their goods at the market price rather than at the price for which they would actually have been willing to part with them. This notion is best illustrated by reference to the supply curve pictured in figure 3.2. This supply curve reminds us that even if the price offered for a particular commodity is very low, someone or few producers would be willing to sell it at that price. And, as the price offered for that good rises, more and more producers are willing to sell greater and greater amounts of it. Let us say for purposes of illustration that some producers would be willing to sell a small quantity of the good represented in figure 3.2 even if the price of that good were as low as P(1) or P(2). Yet, because the supply and demand relationship of this good has produced a market price which is higher, perhaps as high as P(4), these sellers experience a savings equivalent to the difference between the amount they were willing to sell the good for and the amount they were actually able to sell it for in the market. Producer's surplus is the sum total of each of these individual seller's gains. This sum is equal to the area of the lined triangle above the supply curve and below the price line in figure 3.2.

When new technologies are introduced in agriculture, they typically allow a dramatic expansion in the production of the particular farm commodity on which they are used. And as they push the supply curve for this good to the right, they inevitably alter the size of the consumer and producer surpluses associated with it. Researchers attempting to measure empirically the costs and benefits of these changes have argued that their net economic gain or loss to society is approximately equal to the combined changes in the producer and consumer surpluses associated with the commodities affected by these technological changes.

FIGURE 3.3

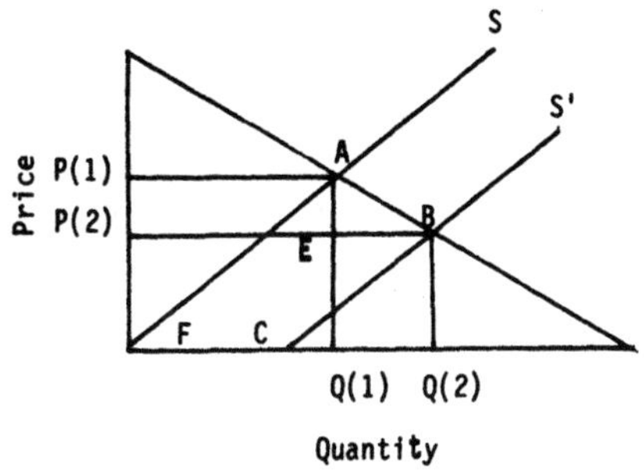

Again, this argument can be most easily seen by reference to the intersecting supply and demand curves pictured in figure 3.3. I have already noted that the size of the consumer's surplus associated with a particular commodity is equal to the area of the shaded triangle under its demand curve and above its price line, and that the size of the producer's surplus associated with that commodity is equal to the area of the shaded triangle above its supply curve and below its price line. Now, when a hypothetical new technology is introduced which shifts the product's supply curve outward from S to S', it lowers the product's price and alters the size of the producer and consumer surpluses associated with it. Whereas the consumer's surplus was equal to the triangle AP(1)D before the technological change referred to in figure 3.3, after that change it was equal to the area of triangle BP(2)D. Or, in other words, the technological change has increased the amount of the consumer surplus by an amount equal to the area BAP(1)P(2). But, not all of this increase in consumer surplus can be seen as a net gain to society from adoption of the new technology because, as you can see in figure 3.3, expansion of the area approximating the consumer surplus reduces the size of the area representing the producer's surplus by an amount equal to the area AEP(1)P(2). The net gain to society from adoption of the new technology, therefore, is represented only by the area within the triangle ABE and the area BCFE, the newly enclosed area above the new supply curve S' and below the new price line. This area, representing the net societal gain from a new agricultural technology, can be estimated mathematically if reliable estimates of the product's elasticity of supply and demand and the percentage shift outward in the supply curve are available. [4]

The SRR approach described above was initially presented by Zvi Griliches in the late 1950s as an effort to measure the returns to research on hybrid corn. [5] It has continued since that time to guide most subsequent efforts to measure major costs and benefits associated with specific new agricultural technologies. In spite of its widespread use, however, this approach has several serious drawbacks which combine to give a very incomplete and misleading picture of the social, economic and political impacts of new agricultural technologies. [6] The most obvious drawback is its failure to indicate how the benefits generated by the new technology are actually distributed among the many groups in society affected by it. We have noted already that mechanization in agriculture has important consequences for a wide range of groups. It affects growers, landowners, farm workers, agricultural credit and transport firms, food processing companies and, of course, businesses which manufacture and sell farm machinery. Yet, as initially formulated, the SRR approach focused only on the effects new agricultural technologies had on "consumers" and "producers." There was no further disaggregation of these broad social groupings which allowed for examination of the different effects mechanization had on such disparate "producers" as displaced farmworkers on one hand, and large growers on the other.

Professors Schmitz and Seckler first attempted to deal with this limitation of the SRR approach in their study of benefits incurred by society as a result of the adoption of the mechanical tomato harvester. Like analysts who had used the approach earlier, they relied on the concepts of consumer and producer surpluses to estimate the social returns generated by this innovation, but unlike earlier analysts, they were careful to distinguish between the "gross social returns" generated by the mechanical harvester and the "net social returns" of this innovation. In computing the latter, they took into account the monetary value of the costs incurred by farmworkers displaced by the harvester. Assuming that displaced tomato workers had no alternative employment opportunities, they estimated the net rate of social return on the harvester to range from 345 percent to -8 percent. More optimistic assumptions about reemployment boosted these estimates of net rate of social return to as high as 1,288 percent. [7] Since Schmitz and Seckler did not attempt to estimate the actual reemployment possibilities of displaced farmworkers, however, they were not able to provide us with the <u>actual</u> rate of return to society from research and development funds expended on the mechanical tomato harvester.

While the study by Schmitz and Seckler undoubtedly advanced the SRR approach by disaggregating the costs incurred by displaced farmworkers from the gains and losses experienced by other "producers," it by no means completely solved the disaggregation problem inherent in this approach. Schmitz and Seckler still aggregated under the "producers" ruberic such disparate groups as growers, retained farmworkers and providers of other farm inputs, without providing a way of assessing the differential effects of mechanization on these different groups. [8] Also, their altered approach did not assess the differential impacts of mechanization on different members <u>within</u> the same groups. We have already noted, for example, that all farmers are not affected the same by mechanization. Typically, larger and more well financed farmers are the first to adopt large-scale mechanical innovations. As a result, they are able to capture a disproportionate share of the surplus profits these innovations generate in the short run. Smaller farmers, on the other hand, often have been unable to afford these innovations. They often have been forced either to abandon farming altogether or move to a less lucrative crop. Remember that, in another study of the mechanical tomato harvester, adoption of this innovation was found to have eliminated 85 percent of the smaller growers in the processing tomato industry. Obviously, then, the innovation studied by Schmitz and Seckler had very different effects on different size farmers. Yet, these inter group distributional consequences were completely masked by the SRR approach to assessing the costs and benefits generated by the mechanical tomato harvester. By masking many of the more important inter and intra group distributional consequences of mechanization, the SRR approach has helped to obscure the unequal way in which the process generates costs and benefits for different elements in society. Because equality is an important value held by many people (the author included), masking mechanization's potential to generate inequality has contributed to an overly optimistic view of this important technological change.

In an effort to overcome the failure of the SRR approach to reveal the distributive consequences generated by new agricultural technologies, researchers most recently utilizing the approach increasingly have tried to disaggregate such broad groupings as "producers" and "consumers" into smaller, more theoretically relevant components. In a recent study of technological change in the Colombian rice industry, for example, Scobie and Posada utilized the traditional

SRR approach to estimate both gross and net social benefits derived from adoption of modern rice varieties. They extended the potential of this approach for distributive analysis, however, by first considering the incidence of research costs incurred by producers and consumers, respectively, as well as by upland and irrigated farms within the producer sector; and second, by examining the distribution of gross benefits, research costs and net benefits across income groups for consumers, and between upland and irrigated farmers for producers. The authors concluded that virtually all of the gains of this new technology were captured by consumers at the expense of producers. But, they also noted that important distributional consequences occurred within, as well as between, these two major sectors. With regard to producers, they concluded that small upland farmers suffered relatively greater losses than other farmers in income foregone due to lower rice prices brought about by the increased supplies attributable to the modern rice varieties. With regard to consumers, the authors concluded that because:

> rice is disproportionately consumed by the lower income groups who make limited tax contributions, the net benefits of the research program were strongly biased toward them in both absolute and positive terms. While the lower 50% of Colombian households received about 15% of the household income, they captured nearly 70% of the net benefits of the research program. [9]

By focusing on important distributive consequences, as well as attempting to estimate total societal gains generated by the modern rice varieties, Scobie and Posada undoubtedly have provided a more complete picture of the social impacts of the new technology than had previous analysts working solely within the SRR approach. Most importantly, they outlined more clearly the trade-offs between heavy costs for small upland farmers and major gains, of both a relative and absolute nature, for the populous lower class urban consumers. Knowing the nature of these trade-offs permits policymakers to make a more informed decision about the desirability of continued or expanded public support for high yield rice varieties.

The study by Scobie and Posada clearly demonstrated that the SRR approach could be used to explore more of the distributional issues associated with new technologies than had previously been done. But, it by no means demonstrated the capacity of this approach to handle all the major distributional issues surrounding the mechanization of agriculture. We noted earlier that increasingly the distribution of benefits generated by technological change depends upon the degree of market power possessed by different actors in the food chain. Yet, neither Scobie and Posada, nor any of the other researchers working solely within the SRR approach, have made an effort to evaluate the market power of actors affected by the technology they have analyzed. In the case of modern rice varieties adopted in Colombia, such an evaluation may not have been necessary. Perhaps the Colombian food industry has not been marked by the growth of economically powerful growers' organizations, food processing and distribution corporations, and unionized work forces. But, as I noted earlier, this clearly is not the case in the United States and most other advanced industrial societies. In these countries the exercise of market power by actors in the food chain is the rule rather than the exception. Moreover, certain of these actors typically hold a great deal more market power than others. An unwillingness to examine this critical variable, therefore, prevents researchers working solely within the framework of a quantitative cost-benefit analysis from pointing out the likelihood that gains generated by mechanization are increasingly usurped by the economically strongest actors in the food chain. It keeps them from emphasizing, for example, that such well-organized groups as grower organizations, food processing and transport firms, and unionized workers are increasingly likely to capture the lion's share of benefits generated by new farm machinery. Failure to emphasize this likelihood helps to further obscure mechanization's potential for exacerbating economic inequality.

It is understandable, of course, why agricultural economists conducting wholly quantitative cost-benefit analysis of new agricultural technologies have shied away from examining the market power of actors affected by these changes. While it is a critically important variable in explaining the actual distribution of technological benefits, market power is not an easy thing to measure quantitatively. Moreover, economists have not yet uncovered a functional mathematical relationship which would have allowed them to link changes in this variable to changes in the distribution of costs and benefits generated by a particular technological change. They have never had, for example, a quantitative procedure which would have allowed them to predict that a measurable increase in the market power of a particular food processor would bring about a measurable decrease in the technological gains enjoyed by a particular group of consumers. Unable to make this quantitative link, they have shied away from even a qualitative judgment concerning the likely effect of increasing market power on the actual distribution of technological benefits. To understand why most Neo-classical economists have shied away from more qualitative assessments of technological change in less competitive market conditions is not to exonerate them for it. Indeed, a primary motivation behind this research is the conviction that more holistic, qualitative assessments may provide us with a better understanding of technological change than quantitative assessments that are misleadingly irrelevant. Regardless of how one feels about this issue, however, it is clear that the purely "objective" stance taken by most Neo-classical economists has helped, if only by default, to obscure some of the least desirable aspects of the mechanization process.

In an effort to provide "objective" cost-benefit analysis of new agricultural technologies, researchers working within the "Social Rate of Return" framework, have papered over mechanization's potential for generating inequality in another way as well. All of these researchers have implicitly assumed that the real losses and gains experienced by different income groups affected by mechanization could be measured by their dollar gains and losses. They have assumed, for example, that a dollar income gain or loss experienced by an impoverished farmworker was equal to the real value of a comparable income gain or loss experienced by a wealthier grower or landowner. It seems at least as reasonable, however, to assume that the law of diminishing marginal utility applies to money as well as other goods. In which case, the real value of an income gain or loss experienced by an impoverished farmworker is greater than the real value of a comparable income gain or loss experienced by a wealthier grower, landowner or consumer. Arguing the validity of this proposition, Montek Ahluwalia and Hollis Chenery of the World Bank some time ago suggested the desirability of using "poverty weights" when evaluating the impact of development programs on the social welfare positions of different income groups in society. For illustrative purposes only, they suggested that income gains of the lowest two quintiles of the population (measured in terms of income) might be assigned a weight of 0.6. The income gains of the next two highest quintiles might be weighted by 0.1. This willingness to weight differently comparable dollar gains and losses experienced by different groups affected by new programs, of course, required these researchers to make a subjective decision. But, it did not require any more of a subjective decision than was already involved in weighting these dollar gains and losses equally. The important point here, however, is that by implicitly utilizing this latter value judgment, researchers conducting quantitative cost-benefit analysis consistently have underestimated the inequalities generated by mechanization in agriculture.

It is not simply the tendency of quantitative cost-benefit analysis to underestimate the impact of mechanization on social equality which has conveyed an overly optimistic assessment of this process, however. Equally important is the fact that these studies have excluded from analysis altogether several important but difficult to measure costs. To date, none of these studies have considered the deleterious impacts that technological change might have on employment, wages and profits in competing commodity industries. In their study of the tomato harvester, for example, Schmitz and Seckler did not consider how the adoption of this innovation would affect the wages of farm laborers working with other crops in the same area. Yet, the depressing effect that mass displacement of tomato pickers would have had on the wages of farm laborers employed in other crops must have been an important cost generated by this new technology. By failing to include the cost, Schmitz and Seckler failed to capture fully the social cost generated by the tomato harvester.

Technological changes in one crop may affect more than just the wages paid in another crop. They also may affect total output and profits in the other crop industry. This is particularly true when consumers can easily substitute the technologically affected product for the other crop with no loss in personal utility. If, for example, consumers can easily substitute apples for oranges without feeling any sense of loss, a technological innovation that sharply reduces the price of apples relative to oranges will generally cause consumers to substitute cheaper apples for the more expensive oranges. As consumer demand for oranges shrinks, so will total employment and profit rates in that crop industry. It is in this sense that technological advances in one crop can generate losses in other competing crop industry. By ignoring these losses, the SRR approach again gives us an overly optimistic picture of technological changes in agriculture.

The costs that mechanization imposes on competing crop industries are not the only ones ignored by researchers working within the SRR framework. We noted earlier that mechanization has played an important role in transforming a nation of largely independent entrepreneurs into a nation of largely servile organization men. It also has been accused of making the food we eat "tasteless." Surely, these are important costs which should not be ignored simply because they cannot be precisely quantified. Yet, this is exactly how current cost-benefit analysis have handled these undesirable by-products of mechanization on the farm.

It is not only the methodological weaknesses characteristic of the economic surplus approach that have contributed to the overly optimistic picture of technological change in American agriculture. The almost exclusive focus of these studies on technological innovations which are widely regarded as "success stories" also has contributed to this distortion. By success stories, I mean those technological innovations which are known to generate major cost of production savings. This emphasis on success story innovations had led analysts to ignore the more common technological changes which generate only modest cost of production savings; savings which are less able to counterbalance social costs they may generate. Part of this bias is due to the over-representation of studies focusing on yield increasing innovations in agriculture, as opposed to capital-intensive labor-saving innovations. The great bulk of literature on this subject suggests that this latter kind of technological change is likely to generate the highest social costs; costs such as rural unemployment, depressed farmworker wages, concentration of land ownership, etc. But, only a small percentage of all these cost-benefit studies have focused on this less promising type of innovation. By under representing technologies expected to have higher than average social costs, the cumulative results of these studies suggest an even more distorted picture of technological change than do the individual studies when viewed separately.

By this point it should be clear that existing studies of agricultural innovations conducted within the SRR framework and relying heavily on the concept of economic surplus present at best an incomplete picture of the costs and benefits generated by continued mechanization. Focusing almost exclusively on success story innovations, and resting upon debatable assumptions about market structure, these studies almost certainly have provided an overly optimistic view of the mechanization process. This does not mean, however, that we should abandon altogether the methodology and concepts used in the SRR approach. But, if we hope to provide a more accurate picture of this kind of technological change, we at least need to: (1) apply this approach to a more representative group of new farm technologies, focusing on economically less promising changes as well as "success story" innovations; (2) supplement the estimates of economic surpluses generated by this approach with an analysis of the market power of major actors in the industry under study and their ability to exercise this power to control the benefits of technological change; and (3) make a greater effort to approximate the actual reemployment possibilities faced by displaced farmworkers. By incorporating these recommendations, the case study presented in chapter 4 hopes to present a more balanced view of the costs and benefits generated by the continued mechanization of American agriculture.

Footnotes

1. The core studies employing the Social Rate of Return approach to technology assessment are: Zvi Griliches, "Research Costs and Social Returns: Hybrid Corn and Related Innovations," Journal of Political Economy, October 1958, pp. 419-431; Willis Peterson, "Return to Poultry Research in the United States," Journal of Farm Economics, August 1967, pp. 656-669; Andrew Schmitz and David Seckler, "Mechanized Agriculture and Social Welfare: The Case of the Tomato Harvester," American Journal of Agricultural Economics, November 1970, pp. 569-577; Yujiro Hayami and Robert W. Herdt, "Market Price Effects of Technological Change on Income Distribution in Semisubsistence Agriculture," American Journal of Agricultural Economics, May 1977, pp. 245-256; Grant M. Scobie and Rafael Posada T., "The Impact of Technical Change on Income Distribution: The Case of Rice in Colombia," American Journal of Agricultural Economics, February 1978, pp. 85-92. Although other analyses have employed the Social Rate of Return framework as a form of technology assessment, these are widely regarded as the most significant studies in this literature. As such, they are the main target of my methodological criticisms.
2. Kenneth Boulding, "The Concept of Economic Surplus," American Journal of Economic Review, December 1945, p. 852.
3. The explanation I provide for the concepts of consumer and producer surplus is taken largely from "The Concept of Economic Surplus and Its Use in Economic Analyses," by John Currie et. al., found in The Economic Journal, December 1971, pp. 741-799.
4. The explanation I provide for the measurement of costs and benefits of technological change by use of the Economic Surplus approach is a simplified version of that presented by Reed Hertford and Andrew Schmitz in Resource Allocation and Productivity in National and International Research, eds. Thomas Arndt et. al. (Minneapolis: University of Minnesota Press, 1977), pp. 148-167.
5. Zvi Griliches, "Research Costs and Social Returns: Hybrid Corn and Related Innovations," op. cit.
6. Several of the drawbacks I note regarding the SRR approach have already been noted by Jurg Bieri et. al., "Agricultural Technology and the Distribution of Welfare Gains," op. cit., American Journal of Agricultural Economics, December 1972, pp. 801-808.
7. Andrew Schmitz and David Seckler, "Mechanization and Social Welfare," op. cit., p. 574.

8. For a thorough discussion of which groups are actually included under the "producer" ruberic and under what circumstances, see Hertford and Schmitz, "Measuring Economic Returns," op. cit., p. 152.
9. Grant Scobie and Rafael Posada T., "The Impact of Technical Change on Income Distribution: The Case of Rice in Colombia," op. cit., pp. 88, 89.
10. Montek S. Ahluwalia and Hollis Chenery, "The Economic Framework," Redistribution with Growth (London: Oxford University Press, 1974), p. 41.

4
The Electronic Tomato Sorter: A Case Study in the Costs and Benefits of Agricultural Mechanization

This chapter is a case study of the costs and benefits generated by the electronic tomato sorter, a new capital intensive technology currently being introduced into the California processing tomato industry. I have focused on the electronic tomato sorter because it is very much at the heart of the political controversy described in chapter 1. Indeed, were it not for the fact that the electronic sorter posed an immediate threat to the livelihood of several thousand farmworkers, it is unlikely that the debate over farm mechanization in general would ever have attained the prominent status it has on the California political agenda. Many of the farmworker protests and legislative hearings on "mechanization" have, in fact, been concerned almost exclusively with the electronic sorter and the dislocations it threatened. Although introduction of the electronic sorter has sparked intense debate over the pros and cons of agricultural mechanization in general, there has been very little careful study of the costs and benefits generated by this particular innovation. This case study, therefore, is first and foremost an effort to contribute to a fair evaluation of a highly controversial new farm technology. As a first step in this direction, I will use recently developed data on the processing tomato industry to calculate estimates of the new technology's producer and consumer surpluses and its "net social product." As I pointed out in chapter 3, these quantitative estimates can be dangerously misleading when the assumption of competitive markets underlying them is not fully met. When properly interpreted, however, they can be a useful empirical benchmark in any comprehensive technology assessment. Calculation of these estimates for the electronic sorter, therefore, provides a useful piece of information that should be included in a thorough evaluation of this controversial new technology. Moreover, because the processing tomato industry is strongly suspected of falling short of the competitive ideal, a case study of the electronic tomato sorter provides an opportunity to explore the extent to which these quantitative estimates may distort the costs and benefits generated by a new technology in a less than perfectly competitive industry. This case study also provides an opportunity to illustrate how some of the quantitative estimates provided by the economic surplus approach can be supplemented usefully by other forms of empirical information to provide a more helpful assessment of new farm technology.

Finally, a case study of the electronic tomato sorter represents an opportunity to study a type of mechanization which occurs regularly in American agriculture, but which has been grossly underrepresented in the existing cost-benefit literature on technological innovation in farming. In chapter 3, I concluded that much of the categorically benign view held by policymakers concerning mechanization was due to the almost exclusive focus of quantitative cost-benefit studies on new farm technologies which are widely regarded as "success stories." By "success stories" I mean innovations which generate obvious and dramatic increases in farm output with no increase in farm inputs, or innovations which maintain existing levels of output with obvious and dramatic decreases in farm inputs. Many technological advances in American agriculture, however, are of a more modest character. They provide for only a slightly greater farm output from a constant level of inputs, or they permit maintenance of existing farm output with only a slightly reduced level of inputs. The electronic tomato sorter exemplifies this common but more modest kind of technological change. It is a highly capital-intensive technology which reduces employment opportunities for a significant number of farmworkers. But, it has no apparent "output effect" and, therefore, appears to generate only minimal alternate employment for those farmworkers it displaces. Moreover, it provides farmers with only very slight savings in their production costs. And finally, it is being introduced into an industry which appears to be characterized by a distinct lack of competition at both the grower and processor levels. All of these factors combine to make the tomato sorter the kind of high risk technological change which has been underrepresented in the cost-benefit literature on this subject. Hopefully, its inclusion in this literature will contribute to a more representative sample of these case studies. This, in turn, should allow us to develop more reliable generalizations regarding the social costs and benefits associated with these innovations.

Before actually launching into the research at hand, I feel that a disclaimer of sorts is appropriate. While I mean this study to contribute to a comprehensive assessment of the electronic tomato sorter, I do not expect that it alone will constitute such a study. Most technological changes generate such a wide range of effects that no single study is likely to provide a truly complete assessment of their costs and benefits. This is certainly the case with the electronic tomato sorter. The newest tomato harvester technology, for example, generates some costs and benefits that can be measured in terms of energy conservation, environmental quality and consumer taste. But, I have not attempted to assess such effects in this study. Nor have I attempted to calculate the research costs involved in developing the electronic tomato sorter so that I could calculate "society's rate of return" on this research investment. Instead, I have focused only on the more limited task of analyzing the impact of this latest tomato technology on consumers' savings, growers' profits and costs, and farmworkers' incomes and working conditions.

TECHNOLOGICAL CHANGE AND TOMATO GROWING

While evaluating the electronic tomato sorter, it is important to keep in mind that this innovation is only the latest in a long series of technological changes adopted by tomato growers in the post World War II period. During this period, tomato growers have moved from hand to mechanical seeding techniques and from hand to mechanical harvesting techniques. Undoubtedly, the most dramatic of these innovations in recent years was the mechanical tomato harvester itself. Before its introduction, processing tomatoes in California were harvested by an army of migrant farm laborers. The vast majority of these migrant laborers were Mexican "braceros" who were allowed into the United States specifically to augment the U.S. agricultural work force. As many as 50,000 of these migrant workers would descend on the tomato growing regions of California during the four to eight week harvest period. There local growers would hire them to move through the fields as many as three or four times at intervals of a week or two to pick the tomatoes as they ripened on the plants. All of this changed dramatically, however, with the introduction of the mechanical tomato harvester in the mid 1960s. The mechanical harvester was a self-propelled piece of farm machinery that moved along a row of tomato plants cutting them off slightly below the soil level by use of large moving blades. By means of a conveyor belt, it then lifted the severed plants onto the harvester where shaking devices separated the tomatoes from the vine. Not all of the tomatoes lifted up onto the harvester were suitable for market, however. A significant percentage were either too green, moldy or damaged to be shipped on to the tomato canneries. These tomatoes had to be sorted out and discarded. This task was performed by a crew of fourteen to sixteen farmworkers known as "sorters." They rode the harvester and worked over the conveyor belts that brought the tomato plants up from the field. As less than marketable produce passed by on the belts, they manually removed it from that which would be sent on to the canners. Whereas in the pre-harvester era as many as 50,000 field hands had been employed to harvest tomatoes, now only 18,000 sorters were needed to handle the same task. [1]

A decade after the original introduction of the mechanical tomato harvester, manufacturers of farm machinery moved on to the next and most recent step in the mechanization of the tomato harvest, the electronic tomato sorter. In 1976, the Blackwelder Farm Machinery Company introduced a mechanical harvester equipped with a color sensitive eye capable of separating red (ripened) from green (unripened) tomatoes. These electronic sorters are mounted over each of the two conveyor belts on the mechanical harvester. As tomatoes are brought up from the field and move along the belts, electronic eyes reject the unripened tomatoes, leaving mainly marketable produce to be sent on to the canners.

In spite of the fact that the electronic sorter adds $40,000 to the price of a new mechanical tomato harvester [2], it has been adopted rapidly by tomato growers. This is primarily because it allows them to reduce sharply the number of farmworkers they must employ as "sorters." Whereas they formerly needed fourteen to sixteen sorters on each mechanical harvester, use of the electronic sorter allows them to reduce this number to seven or eight. [3] There are times during the harvest season, however, when even the electronically equipped tomato harvesters require a larger crew. This situation arises from the limitations of the electronic sorter itself. In its present state of development, the new technology cannot be used very efficiently in the later stages of the tomato harvest. In these stages, the major sorting function becomes one of separating ripe tomatoes affected by mold, sunscald and other maladies, from ripe tomatoes suitable for market. Because the electronic sorter is capable only of separating ripened from unripened tomatoes, growers must compensate for the new technology's limitations by increasing the number of manual sorters working on the harvester. In spite of this limitation, however, agricultural extension agents at the University of California have estimated that electronically equipped harvesters are able to save growers seventy-two cents per ton of harvested tomatoes. [4]

THE COSTS AND BENEFITS OF ELECTRONIC SORTING: AN ECONOMIC SURPLUS APPROACH

What are the costs and benefits associated with the move from hand to electronic sorting in tomato harvesting? How have these costs and benefits been distributed among the major groups in society affected by this new technology? The traditional approach to providing wholly quantitative responses to these questions has relied on the concept of economic surplus discussed in chapter 3. Keeping in mind the various shortcomings in this approach, I have calculated that, in the nine major tomato producing counties in California, the electronic sorter generates annually a $10,311,346 gain in consumer surplus. It also generates annually a $6,005,175 loss in producer surplus. [5] Combining these two surpluses yields an annual "gross social product" of $4,306,171. [6] As has been pointed out already by those using the economic surplus approach, however, it is only after the labor displacement costs generated by the electronic sorter have been subtracted from the "gross social product" that we can estimate a new technology's "net social benefit" (nsb). Although noting the need to reduce the gross social product by the cost of labor displacement, practitioners of the economic surplus approach typically have not attempted to estimate the actual labor displacement costs generated by the technologies they study. Instead, they have limited themselves to presenting alternative assumptions about farmworker reemployment rates, and then calculating labor displacement costs based on these alternate assumptions. As shown in Table 4.1, following this practice with the electronic sorter indicates that if all those displaced find alternate employment, then the labor displacement cost of this new technology is zero. And, the sorter's annual "net social benefit" is equal to its annual "gross social product" or $4,306,171. If only 75 percent of those displaced are reemployed, then labor displacement costs for the electronic sorter are estimated at $1,373,829. [7] Subtracting this figure from the gross social product of $4,306,171 leaves a net social benefit of $2,932,342. If only 50 percent of those displaced find alternate employment, then the net social benefit of the sorter falls to $1,558,513

TABLE 4.1
Farmworker Reemployment Possibilities, Labor Displacement Costs, and Net Social Benefit (NSB) Estimates for Electronic Sorter

Percent of Displaced Farmworkers Reemployed	Labor Displacement Cost	Estimates of NSB
100%	-0-	$4,306,171
75%	$1,373,829	$2,932,342
50%	$2,747,658	$1,558,513
25%	$4,121,487	$ 184,684
0%	$5,495,316	$1,189,145

annually. Finally, of course, if none of those displaced by the new technology find alternate employment, then the labor displacement costs of the new technology will rise so high that the electronic sorter will generate a negative "net social benefit" of -$1,189,145 annually.

A CLOSER LOOK

While estimates of the electronic sorter's economic surpluses and gross social product are useful indicators of important costs and benefits generated by this new technology, they clearly do not tell us all we need to know in order to assess its social worth. In the first place, they tell us only that the technology's "net social benefit" ranges from a low of -$1,189,145 to a high of $4,306,171 annually, depending on the reemployment possibilities of displaced farmworkers. They do not tell us, however, what these reemployment possibilities are likely to be. Moreover, as I noted in chapter 3, the estimates of economic surpluses themselves hinge on questionable assumptions about the market structure of the processing tomato industry. To know if these estimates are valid, we must examine the economic actors in this industry more closely. Specifically, we need to know if any of these actors possess sufficient market power to usurp whatever benefits might be generated by the electronic sorter. In an effort to get this required "closer look" at major actors in the processing tomato industry, I have turned first to two surveys sponsored by the University of California, Davis, in late 1977. [8] The first of these was a random sample survey of sixty tomato growers Ann Scheuring conducted in two major tomato producing counties in California. The second of these was a survey of eighty tomato sorters conducted by a team of bilingual interviewers on randomly selected tomato farms in Yolo County, California. I have attempted to supplement these surveys with data available in a number of published studies and documents dealing with the California processing tomato industry. I believe these additional sources of information augment the quantitative estimates generated by the economic surplus approach and contribute to a more informed assessment of the electronic tomato sorter.

ELECTRONIC TOMATO SORTING AND DISPLACED FARMWORKERS

It is clear from estimates provided by the economic surplus approach that the electronic sorter's "net social benefit" depends largely on the reemployment possibilities of displaced farmworkers. If a significant percentage of these displaced tomato sorters find alternate employment, this new technology may provide a net benefit to society. If, on the other hand, most of those displaced remain unemployed, the new technology will clearly generate a net social cost. Unfortunately, there is no way to estimate precisely what the long-term reemployment possibilities of this critical group will be. This is largely because we have no way of knowing how the many factors affecting these possibilities will change in the distant future. We can, however, use the information available on this work force to make a rough estimate of the number of displaced tomato sorters who are likely to find alternate employment in the near future given existing economic conditions.

Judging from other studies of displaced and unemployed workers, the most important factors affecting the short-term reemployment possibilities of displaced tomato sorters are their: (1) educational backgrounds, (2) alternate job skills, and (3) youth. [9] Presumably, the first of these factors also encompasses the ability of displaced farmworkers to speak English. It was with an eye towards collecting information on these employment related personal characteristics that the University of California sponsored a series of 80 interviews with tomato sorters in 1977-78. According to the information obtained in these interviews, most tomato sorters will find it extremely difficult to get alternate employment once they are displaced by the electronic sorter, and they will find it doubly difficult to find alternate non-farm employment. Only 19 percent of the tomato sorters we interviewed, for example, spoke English. The remaining 81 percent spoke only Spanish. The average number of school years completed by these workers was six. Only 22 percent have had some education beyond the ninth grade. And, only one individual of the 80 sorters interviewed had any schooling beyond the high school level. No more than 20 percent of those we spoke with had any employment experience outside of farm work during the previous two-year period.

Apparently, the most important personal resource that tomato sorters have to aid them in their reemployment efforts is their youth. The median age of tomato sorters interviewed was 30 years old. Only 13 percent were 45 years or older. Unfortunately, however, only a very small portion of tomato sorters have the optimal combination of youth, education and facility with English needed to give them a high probability of finding alternate, non-farm employment. As you can see from the cross tabulation of sorters' age and education in Table 4.2, only 19 percent of the tomato sorters in Yolo County were below 30 years old

TABLE 4.2
Age and Education Level of Yolo County Tomato Sorters

	Education		
Age	Grades 1-6	Grades 7-9	Grades 10 and above
15-30	19 23.7%	6 7.5%	15 18.8%
31-40	21 26.2%	6 7.5%	2 2.5%
41 and over	10 12.5%	0 0.0%	0 0.0%

and had some high school education. It is also this group of sorters who were most likely to speak English. Presumably, these individuals have a reasonable opportunity of finding alternate non-farm employment as tomato sorting work evaporates. A majority of tomato sorters, however, face much dimmer reemployment possibilities. Sixty-two percent of those interviewed were either more than 40 years old or had less than a seventh grade education. Moreover, 90 percent of this group spoke no English. These sorters, no doubt, will find it difficult, at least in the near future, to move to other forms of employment as work in the tomato harvest declines. This is particularly true with regard to non-farm employment.

I have drawn the above picture of farmworkers reemployment possibilities by relying on the work related personal characteristics of tomato sorters in Yolo County. A similar picture was suggested by tomato sorters themselves when they were asked about alternative employment. Forty-six percent of the tomato sorters interviewed indicated that if they had not been working as sorters, they would have had no other form of employment available to them. Another 24 percent felt that their only alternative form of employment would have been other farm work. Together, these figures indicate that 70 percent of those tomato sorters interviewed in Yolo County see themselves as unable to obtain non-farm reemployment as their work in the tomato harvest declines. Whether we rely on the subjective assessments of farmworkers themselves (54 percent believed they could find alternate employment), or on a more objective review of the work related personal characteristics of tomato sorters (19 percent had the youth, education and language abilities to make them good reemployment risks), the conclusion seems fairly clear and consistent. At best, slightly more than half of the tomato sorters displaced by the electronic sorter can reasonably be expected to find alternate employment as their work in the tomato harvest is eliminated.

Having a clearer idea about short-term farmworker reemployment probabilities, we are now able to make a more accurate estimate of the annual losses tomato sorters will incur in the near future as growers shift to more capital-intensive harvest techniques. We noted earlier that complete adoption of the electronic sorter could cost displaced farmworkers anywhere from zero to $5,495,316 per year depending upon their reemployment possibilities. We have just estimated, however, that at least in the very near future, no more than 19 to 54 percent of those displaced are likely to find alternate employment. Even assuming that the higher of these two reemployment figures is reached, the actual value of farmworker wage losses induced by the electronic sorter will amount to $2,527,845 annually. If we subtract this cost from the value of the "gross social product" ($4,306,171) estimated earlier, we see that the maximum annual net benefit this new technology may generate for society in the short term is $1,778,326 annually.

But, should we simply subtract the dollar value of the farmworkers' lost wages from the gross social product (gsp) to find the electronic sorter's net benefit to society? I pointed out in chapter 3 that critics of the economic surplus approach have questioned the implicit practice of weighting equally the dollar gains and losses of groups from different income levels. Yet, this is essentially what I have done above by subtracting the dollar value of farmworkers' lost wages from the gsp (remember, the gsp is simply the aggregate of consumer and producer surpluses). Critics of this practice have argued that it ignores the fact that a dollar loss is dearer to a poor man than it is to a rich man. If this criticism is accurate, it suggests that we may need to weight the wage loss of farmworkers more heavily than the losses or gains of producers and consumers affected by the electronic sorter. This additional weighting of farmworker wage losses would only be justified, of course, if the income level of this group was significantly lower than that characteristic of these latter two affected groups.

Are the farmworkers who are being displaced by electronic sorters actually poorer than the consumers who might benefit from this new technology? Again, the best possible sources of information on this question are the interviews we conducted with tomato sorters in 1977. According to these interviews, the average annual income of tomato sorters in Yolo County earned from all sources of employment in 1976 was $2,852. And, between January 1st and October 10th of 1977, the date the interviews were conducted, these same tomato sorters had received from all their income sources, including unemployment insurance and food stamp benefits, an average of only $3,039. During this same period, the average income earned by all workers in the United States was $6,441. [10] A similar picture of relative farmworker poverty emerges if one looks at family income rather than individual income. In 1976, the average family income of tomato sorters interviewed in Yolo County was $9,352. Through October of 1977, the family incomes of sorters averaged $10,748, including benefits received from unemployment insurance, welfare assistance and the U.S.D.A. food stamp program. The average family income for the country as a whole in these same years was $14,958 and $16,009, respectively. [11]

The annual incomes of tomato sorters' families are considerably below the national average in spite of the fact that there are an average of 2.4 adults employed in each of the families. Because the small incomes reported by the average sorter must be used to support large families (the average family size reported was 5.7 members), a significant portion of these families live below, or only slightly above, the poverty levels defined by the Social Security Administration of the federal government. Based on their reported family incomes for 1976, 30 percent of the families of tomato sorters were living below the poverty level, and 37 percent were living within only $1,000 of this poverty level. [12]

The above figures on farmworker income suggest that the tomato sorters who are being displaced by the electronic sorter are significantly poorer than the consumers that might benefit from this new technology. I say "suggest" because I do not have figures indicating the average annual individual or family income for those who consume processed tomato products. These products are extremely diverse, however, ranging from whole tomatoes to tomato sauce to common everyday catsup. It, therefore, seems reasonable to assume that they are consumed in one form or another by a wide segment of the entire population. If this is the case, then the average annual income of tomato product consumers is likely to be nearly identical to the average income for the population as a whole. I already have noted above that this average family income for the population was $14,958 in 1976. This was 60 percent higher than the comparable figure for tomato sorters in that same year.

If one feels that a dollar loss is dearer to a poor man than it is to a rich man, then he no doubt believes that wage losses suffered by farmworkers need to be weighted more heavily than the consumer gains generated by the new tomato technology. This, of course, is a value judgment. But, it is no more of a value judgment than the decision to weight equally the gains and losses of consumers and displaced farmworkers. How much additional weighting of farmworkers' wage losses is reasonable? Obviously, there is no objective answer to this question. Any such weighting must be wholly subjective. I, therefore, do not intend to defend any particular weighting scheme. I believe it is important to point out, however, that the wage losses likely to be suffered by displaced farmworkers ($2,527,854) need to be weighted only by a factor of 1.7 to cancel out completely the gross social product credited to the electronic sorter by the economic surplus approach.

Farmworker poverty is not the only factor that needs to be considered when deciding on how to weight losses generated by the electronic sorter. Equally important to many people is the residency status of the farmworkers being displaced. Feeling that the vast majority of those displaced are migrant workers from Mexico, one may decide to discount heavily the labor displacement costs associated with this process. Unfortunately, accurate information on this subject is extremely hard to come by. The vast majority of tomato sorters we inteviewed in Yolo County were people of Mexican origin. Many of these were either United States' citizens or legal aliens who resented being questioned about their citizenship simply because they were Mexican-Americans. Others were quite likely illegal aliens who could hardly have been expected to identify themselves as such to an unknown interviewer. Given these obstacles, we decided to ask tomato sorters a less threatening question about their residency status and hoped that we, therefore, might get more reliable information on this crucial issue. Very near the end of the interview we asked farmworkers "Have you lived anywhere other than Yolo County during the past year?" If they answered yes, we then asked them "Where?" and "How long?" Judging from the responses we received from this question, the tomato sorters facing technological displacement due to adoption of the electronic sorter are primarily local residents. The exact percentage that can be classified as local residents varies, of course, according to the different ways one defines a local and migrant worker. But, 77 percent of the farmworkers we interviewed in Yolo County lived within fifty miles of their work site for at least ten months of this past year. Another 10 percent of those we interviewed lived within fifty miles of their work site for at least six months of the year, but returned to Mexico for between two and six months of the year. An additional 5 percent lived within fifty miles of their work site for at least six months of the year, but returned to another area of California or some other state in the U.S. for between two and six months of this past year. The remaining 8 percent lived in Mexico for more than six months of this past year. If one takes a liberal approach to the definitional problem and considers all farmworkers who have lived within the local area for at least six months of the past year as local residents, fully 92 percent of all sorters employed in Yolo County at the peak of the tomato harvest can be described as local residents. If one takes a more conservative approach to the data, and discounts those farmworkers who lived in the local area for at least six months of the year, but returned either to Mexico or to some other area of the United States for between two and six months of this past year, then only 77 percent of the sorters in the county can be considered local residents. A final approach to determining the residency status of the tomato sorters employed in Yolo County is to temper the objective criteria of residency with farmworkers' self appraisals of their residency status. This approach indicates that 67 percent of the sorters we interviewed lived in the local area for at least six months of the past year and identified a local community as their permanent residence. Regardless of one's interpretation of the data, the conclusion regarding the residency status of sorters facing displacement due to adoption of the electronic sorter by growers in Yolo County is quite clear. The vast majority of these workers are local residents. The precise size of this majority is estimated with a considerable degree of certainty to be at least 66 percent of the work force, and with increasing uncertainty to be as high as 92 percent of the work force. This means that even those analysts using residency status as a relevant weighting criteria must be careful not to discount totally labor displacement costs under the mistaken belief that those losing their jobs are primarily "foreigners." The findings on tomato sorters' residency status have another implication as well. They make it clear that the wage losses suffered by this group are likely to have significant "ripple effects" in the communities where the new tomato harvest technology is being adopted. Unfortunately, at present there does not seem to be any way of estimating the precise magnitude of these ripple effects.

To this point, I have discussed only the losses that the electronic tomato sorter will generate for displaced farmworkers in the "short term." How can we expect the value of these losses to change in the long run? The consensus of opinion among agricultural economists who study this issue is that the value of these losses will necessarily decline and ultimately disappear in the long term. [13] They reason that as time goes by more and more of those displaced by the new farm technology will be forced into alternate employment. Many tomato sorters, for example, who do not speak English when they are displaced by the electronic sorter will learn it. Also, many who presently do not have either the education or job skills to obtain alternate employment will get them in the long run. Finally, regardless of whether they find alternate employment or not, all those displaced by the electronic sorter ultimately will die. At this point, so reason the proponents of this perspective, the value of lost farmworker wages will fall to zero.

Unfortunately, this reasoning is a bit too optimistic. Most importantly it ignores the costs that unemployment or further impoverishment of a tomato sorter impose on his children. Research on the issue of poverty makes it painfully clear that the current generation of poor people in this country often are poor primarily because the previous generation was unable to provide them with opportunities and incentives to overcome this condition. [14] Of course, I have no way to calculate the future cost of poverty and its related anti-social behavior that can be laid at the doorstep of the labor displacing electronic tomato sorter. But, it seems to me that until someone devises a way to measure these costs, any conclusion about the long-term growth or decline of labor displacement costs generated by the tomato sorter must be seen as pure conjecture.

Adoption of the electronic sorter affects not only those tomato sorters who are displaced by this new technology. It also affects the farmworkers who retain their jobs after its adoption. One of the primary benefits a new farm technology often brings retained farmworkers is an improved working situation. Unfortunately, the new tomato harvest technology has not produced this desired effect, at least if we judge from the responses of farmworkers who have worked with it. Among the tomato sorters we interviewed in Yolo County, only 17 percent preferred working on electronically equipped tomato harvesters. Fifty-one percent preferred harvesters where the sorting was done by hand. Another 30 percent expressed no preference at all. Those sorters who disliked working on the electronically equipped harvesters indicated that their dislike of the newer technology resulted from the fact that these machines moved through the fields more rapidly than older machines and, thus, caused them to work faster than they had been used to. They also pointed out that in spite of increased productivitiy, they typically received the same hourly wage as sorters employed on machines where sorting was still done by hand. This last observation appears to clash with the expectations of neo-classical economists who traditionally have predicted that increases in worker productivity would be accompanied by higher wages. This has not occurred in the case of the electronic sorter apparently because the supply of farmworkers is perfectly elastic for current levels of grower production. This is due in part at least to the fact that labor displacement precipitated by the new harvest technology has created an excess supply of hand sorters in relationship to the demand for them. In this situation, farmworkers are unable to capitalize on their increased productivity unless they develop sufficient organizational strength to constrict the excess supply of farm labor. It was in an effort to provide this organizational strength that the United

Farmworkers of America (UFW) attempted to organize tomato sorters in the mid-1970s. This organizational effort met with only limited success, however. And today, tomato sorters in California continue to bargain individually with growers over pay and working conditions.

ELECTRONIC TOMATO SORTING, TOMATO PROCESSORS AND TOMATO GROWERS

The estimates of consumer and producer surpluses and gross social product that I provided for the electronic tomato sorter assume that this technological advance has occurred in a largely competitive industry. The fact that thousands of tomato sorters still continue to bargain individually with growers over pay indicates that at least this industry actor has not yet acquired sufficient organizational strength or "market power" to alter the crucial competitive assumption. But, can we also say the same about tomato growers and processors, the other two major actors in this industry? Apparently not. As long ago as 1959, analysts of the processing tomato industry described it as "oligopsonistic--few firms buying from many essentially unorganized growers." [15] At this time there were over 4,000 tomato growers in California, but only 56 processing firms that bought their tomatoes. [16] In this situation, individual growers had to compete with each other for contracts with major processing firms (H. J. Heinz, Campbell Soup Company, Contadina Foods, Del Monte Corporation, and Hunt-Wesson Foods). Playing growers off against each other, the processors were able to bid down the price of tomatoes to a point where growers were forced to pass on nearly any savings in the cost of production that were generated by more efficient technology.

Tomato processors, of course, would have been forced to pass these technological savings on to consumers if their ranks had been characterized by as much competition as existed between growers. Some agricultural economists studying the industry at that time suggested that this was, in fact, the case. They argued that the industry's price setting practices most closely approximated the "competitive barometric model." [17] In this kind of situation, one of the larger firms in the industry acts as a "price leader," setting a price it thinks reflects the market conditions for the industry as a whole. Other firms in the industry quickly follow suit, unless they believe the "price leader" has miscalculated seriously the market conditions for the industry as a whole. If other firms believe the price to be too low or too high, a new price leader emerges and sets another price. [18]

Not everyone, however, was certain that tomato processors were so competitive. While there were fifty-six processing firms buying tomatoes in California in 1956, the largest eight of these purchased over half of all tomatoes sold in the state. [19] This degree of concentration provided an attractive opportunity for the largest firms in the industry to cooperate in setting a price which guaranteed them above-normal profits. Some indication of this cooperative pricing effort was seen in the fact that each year the major canners paid all growers a uniform price for their tomatoes. Growers and tomato processors did continue, however, to haggle over various non-price provisions in their contracts, and it was this haggling that led some economists to characterize the processor ranks as "competitive" in the mid-1950s. [20]

Whether or not major tomato processors actually engaged in cooperative pricing practices during the 1950s remains unclear. What is clear is that the economic concentration that gave rise to these suspicions has grown even worse since then. Between 1956 and 1975, the number of tomato processing firms in California dropped from fifty-six to twenty-eight. [21] Moreover, the six largest firms in the industry now purchase approximately 80 percent of all tomatoes sold to private firms in California. [22] Finally, there is increasing evidence that the major tomato processors have actually engaged in cooperative pricing practices. This evidence is laid out most clearly in a federal anti-trust suit filed against Heinz and 24 other canners by a tomato grower in 1975. According to the brief filed by Enrico Farms, the tomato grower in this case, tomato pricers from the major canners continually meet with each other to discuss and set a single industry-wide price for processed tomatoes. [23] His suit specifically charges, for example, that tomato pricers from the major canneries had 17 meetings between January 6 and December 16, 1971. Then:

> On January 11, 1972, Borden announced a $28-per-ton price. Hunt-Wesson did the same on January 17, Kern and Heinz on the 19th, Campbell, Valley, Del Monte and Tillie Lewis on the 24th, Cantadina on the 25th, NCC on the 28th, American Home on the 31st, Escalon on February 1 and Libby on the 10th. All offered growers $28-per-ton contracts. [24]

Enrico's suit also alleges that a similar situation occurred the following year. He contends that on

November 20, 1972, William Poggetto of National Can Company had lunch with two tomato pricers from Contadina (Poggetto's own expense records identify the purpose of the meeting as "tomato pricing for 1973") and that, immediately after that meeting:

> Contadina started to contract at $30 per ton. On December 5, 1972, Libby followed suit. Del Monte did the same on December 12, Heinz on the 14th and Kern on the 17th. [25]

In addition to the two examples noted here, the Enrico brief alleges several other specific instances of price fixing by major canners between 1970 and 1975. [26] Whether or not Enrico's lawyers will ever successfully prove to the satisfaction of the court that major tomato processing firms have engaged in price fixing behavior is not yet clear. Although filed in 1975, the suit still is in its preliminary stages, and no final verdict is expected soon. Regardless of how this suit is decided, however, the fact that it documents regular meetings between tomato pricers of major processing firms prior to establishment of uniform contract prices should weaken one's faith in the assumption that this industry is truly competitive. It seems at least as reasonable to assume that in recent years tomato processors have developed sufficient market power to retain at least some of the gains generated by technological advances in tomato farming. Of course, the processors may choose to pass on all of these technological gains to consumers in the form of lower prices. But, they also may decide to translate them into increased profits for themselves or increased wages for their workers. The actual distribution of these gains depends ultimately on the relative bargaining power of the actors involved in the processing tomato industry.

Recognizing this essential fact, tomato growers, the third major actor in the industry, have struggled for many years to organize themselves into a single effective price bargaining unit. After several failed attempts, growers first accomplished this objective in 1974. That year a majority of tomato growers in California refrained from negotiating individual contracts with the canners. Instead, they designated the California Tomato Growers Association (CTGA) as their bargaining agent. Having designated the CTGA as their bargaining agent, growers were then legally prohibited from selling to canners for a price lower than that negotiated by the CTGA. [27] With the advent of successful unit price bargaining by the CTGA in 1974, growers also appeared temporarily to gain the organization necessary to hold on to part of the benefits generated by technological advances in tomato farming. The best indication of this newfound bargaining power was the higher average prices and profits growers received for their produce in the years immediately following the CTGA's first successful bargaining effort. As you can see from data in Table 4.3, both the average price per ton growers receive from tomato canners, and their average net return per acre have always varied greatly from one year to the next. In years when canners perceive a relatively heavy consumer demand for processed tomato products, they offer growers a relatively high price per ton to encourage them to grow enough tomatoes to meet this demand. Conversely, when canners perceive a relatively slack consumer demand, they offer growers a lower price per ton because they don't need as many tomatoes. Variation in consumer demand explains only part of the variation in price that growers receive from canners, however. A second major factor concerns the stock of processed tomato goods that canners carry over from one year to another. If, in any particular year, canners seriously overestimate consumer demand, for example, they are left with a large inventory that must be carried over to the following year. This, in turn, means that canners will need less acreage from growers. They therefore will offer growers a lower price per ton than if they hadn't been stuck with such a large carryover inventory. Because the prices that growers receive vary so much from year to year, it is not surprising that their net return per acre should also vary sharply from year to year. But, the price per ton a grower receives is not the only factor that affects his net return. His yields and production costs also affect this return. Since the former, in particular, can fluctuate sharply from year to year, even a grower who receives a good price per ton from the canner can suffer a poor net return per acre.

In spite of annual variations in price and profit discussed above, the data in Table 4.3 can still be used to shed light on the organizational strength of tomato growers if we look at the average yearly prices and profits received by growers over the course of several years. As you can see from Table 4.3, between 1956 and 1973 growers received an average yearly price per ton of $28.29 (in constant 1967 dollars) for their tomatoes. But, during the four years immediately following the first successful price bargaining effort by the CTGA, that figure rose 16 percent to $32.87. This is the highest average price growers had ever received during any four-year period in the industry's history, except the unrepresentative 1965-1968 period. During these years, growers were switching from hand to mechanical harvesters. And, canners, fearing that if

TABLE 4.3
Summary of Annual Grower Prices and Profits, 1956-1977
(In Constant 1967 Dollars)

Year	Price/Ton Received by Growers	Average Yearly Price/Ton	Growers' Net Return Per Acre	Average Yearly Return Per Acre
1956	$27.88		$ 88.35	
1957	25.97		62.38	
1958	26.21		58.00	
1959	24.97		26.63	
1960	26.38		31.01	
1961	33.59		129.79	
1962	30.46		78.54	
1963	27.69		20.21	
1964	27.23	$28.29	4.29	$72.96
1965	37.46		170.80	
1966	30.86		89.95	
1967	38.70		212.21	
1968	33.78		203.86	
1969	24.77		22.14	
1970	21.67		18.00	
1971	23.08		17.97	
1972	22.35		5.45	
1973	26.29		73.72	
1974	38.46		377.38	
1975	34.49	$32.87	291.84	$235.73
1976	27.80		79.42	
1977	30.74		194.26	

Source: The tomato prices in Table 4.3 come from the California Crop and Livestock Reporting Service. They are the average state-wide price received by growers.

The figures on growers' net returns per acre between 1956 and 1975 are provided in Jon Brandt, An Economic Analysis of the Processing Tomato Industry, unpublished Ph.D. dissertation, Department of Agricultural Economics, University of California, 1977, p. 72. Using the computational formula provided by Brandt, I have extended these figures forward to 1977.

the new technology did not succeed, the entire industry would collapse, were extremely generous with growers. [28] Once the succcess of the new technology was assured, however, canners returned to their more normal parsimonious pricing policies.

Not surprisingly, the higher prices won for growers by the CTGA in the 1974-1977 period also brought them greater profits. As you can see from the data on grower profits presented in Table 4.3, during the four years immediately following 1974, growers' net returns per acre in constant 1967 dollars averaged $235.73 per year. In the eighteen year period prior to 1974, growers' net returns per acre averaged only $72.96 per year. In other words, since 1974 growers averaged 323 percent greater net returns per acre than they did without the help of collective price bargaining efforts by the CTGA. Apparently, a significant part of these new-found benefits can be traced to the fact that tomato growers, like tomato processors, had acquired sufficient market power to retain a share of savings generated by technological advance in their industry.

Unfortunately for growers, the organizational strength they displayed in the mid-1970s appears to have been short-lived. In 1975, 80 percent of all tomato growers in California permitted the CTGA to represent them in contract negotiations with the major canners. [29] These growers controlled 70 percent of all the tomato acreage planted in the state. [30] By 1981, however, only about 35 percent of the tomato growers in the state belonged to the CTGA, and these growers accounted for only about 35 percent of the tomatoes sold to private tomato canners. [31] Obviously, the bargaining strength of the CTGA had eroded drastically in just five years. No doubt this is due in large part to the CTGA's inability to impose production limits on its members. Unable to limit members' output, the CTGA was unable to protect growers from a situation of excess supply in their industry. This situation, in turn, made it increasingly difficult for the CTGA to maintain its role as an effective bargaining agent for growers. Individual growers, anxious to sell their excess produce, increasingly began contracting individually with the major canners. By 1980, the situation had become so bad that the CTGA was unable to negotiate any acceptable contract price with the major canners and, ultimately, released its membership to sign contracts as free agents. [32]

Some analysts may point to this last occurrence in particular as evidence that the market power of the CTGA was always more fiction than fact. They might argue that short-term, excessive consumer demand rather than the market power of the CTGA was behind the higher than average prices and profits growers enjoyed in the mid-1970s. No doubt strong consumer demand and minimal canner carryover stock created the climate for a successful CTGA bargaining effort in 1974. But to suggest that growers had not temporarily gained a significant measure of market power in the mid-1970s is simply misleading. Should we now deny that OPEC exercised significant market power in the mid-1970s, simply because that power has been seriously eroded by the recent glut of oil in the world market? Of course not! Likewise, we should not deny that tomato growers in the mid-1970s had gained sufficient market power to retain an increasing part of savings generated by technological advances in their industry. After all, spokesmen for major tomato processors themselves admitted the influence of the CTGA in obtaining higher prices for growers in these years. [33]

Once growers had attained the organizational strength to retain a share of savings generated by new technology, they apparently left tomato processors with an unpleasant choice. The latter could either cut their own profits or increase the prices consumers pay for processed tomato products. Apparently, processors chose the latter of these two options. As you can see from the data in Table 4.4, between 1974 and 1977 (the last year this item made up part of the CPI Food Index), the market price of canned tomatoes increased a

TABLE 4.4
CPI Data for Tomatoes, and Processed Fruits and Vegetables, 1967-1977 (1967 = 100)

Year	Canned Tomatoes	Processed Fruits and Vegetables
1967	100.0	100.0
1968	104.5	105.6
1969	100.8	106.5
1970	109.0	109.2
1971	115.6	116.2
1972	116.2	120.5
1973	125.2	130.2
1974	149.5	170.6
1975	176.3	178.3
1976	178.3	183.0
1977	195.2	188.8

Source: CPI Detailed Reports, U.S. Department of Labor, Bureau of Labor Statistics, 1967-1977.

significant 45.7 percent. This means that the price of canned tomatoes was increasing 15.2 percent per year during the years the CTGA was at the height of its organizational strength. In the four year period before 1974, the comparable rate of increase in the price of this product was only 5.4 percent per year.

Of course, there is no way to establish exactly how much the successful price bargaining efforts of the CTGA contributed to the higher prices consumers had to pay for canned tomato products between 1974 and 1977. Other, more general inflationary pressures also must have contributed to these price increases. Some indication of the strength of these pressures is suggested by the fact that during this same period the price of all processed fruits and vegetables increased by 18.2 percent. This price increase, however, falls far short of the 45.7 percent increase recorded for canned tomatoes. It, therefore, indicates that forces other than the normal inflationary ones just mentioned were behind the rapid price increase of canned tomatoes. It seems reasonable to conclude that the newly acquired market power of the CTGA was one of these forces.

We have now looked more closely at each of the major actors affected by the electronic tomato sorter. In light of this closer look, how realistic does the assumption of perfect competition which underlies our initial estimates of this technology's economic surpluses appear? If this assumption is unrealistic, how does it change our interpretation of these estimates? In no industry, of course, does one find "perfect competition." The term itself refers to an ideal which is only approximated to a greater or lesser degree in any real situation. But, in the years immediately following introduction of the electronic sorter, the processing tomato industry apparently did not approximate this ideal very closely. Throughout this period, tomato processors seem to have had the market power necessary to retain some of the savings this new technology would have generated for consumers in a more clearly competitive situation. Tomato growers as well had temporarily acquired sufficient market power to retain some of the gains that the new technology might have generated for consumers. Indeed, the history of the processing tomato industry since the mid-1970s can be seen primarily as a struggle between these two relatively well-organized economic actors over the benefits of present and future technological advances. This is manifested most clearly in the plethora of law suits that have been filed by these two actors against each other since 1975. I have already mentioned the Enrico suit filed by a grower against major processing companies. Since then, Garden Valley

Foods, a small canner, has filed suit against the CTGA, [34] alleging that this organization has unlawfully fixed prices among growers. The CTGA, in turn, filed two separate law suits against major canners alleging that the latter have unlawfully discriminated against CTGA members in letting contracts. [35] These law suits, as well as our own analysis, strongly suggest that consumer gains we credited to the electronic sorter based on the assumption of perfect competition are overstated. By how much they are overstated is not clear. Unfortunately, there are no quantitative procedures presently available which allow us to measure "market power" exercised by tomato processors and growers, and then calculate the resultant drain this power puts on a technologically generated consumer surplus. Nevertheless, it seems reasonable to conclude that the electronic sorter generates something less than the $10,311,346 attributed to it earlier by the economic surplus approach. Precisely how much less we simply do not know. It may be 75 percent of the benefit indicated by the economic surplus approach. Or, it may be 50 percent. It may even be less than this. We simply do not know.

If the economic surplus approach has overestimated the benefits that consumers are likely to enjoy from the electronic sorter, it also has blurred our understanding of the costs and benefits producers enjoy from this new technology. In part, this is because the approach has ignored some important long-term gains larger growers, in particular, are likely to reap from the electronic sorter. The most important of these is their reduced vulnerability to union organizing efforts. All growers want to avoid what they see as the disruption in their operations resulting from the imminent unionization of California farmworkers by the United Farmworkers (UFW). The smaller crews required by electronically equipped harvesters allow growers to reduce the size of their labor force and, thus, their vulnerability to union organization efforts. This is a particularly attractive side benefit of the new harvest technology for very large growers, because it is precisely these largest growers who are the primary targets of UFW efforts. Some concrete indication that larger growers are more motivated to switch to the new technology because of the threat of unionization is indicated by information gathered in the University of California sponsored interviews with tomato growers on this issue. One of the questions asked of growers in this interview was "What was your primary motivation for adopting the electronic sorter?" Subsequent analysis of growers' responses to this question indicated that larger growers listed "union threat" as their primary motivation for adopting the new harvest technology more often than their smaller counterparts. Among growers we interviewed in Yolo County, for example, the median acreage grown by farmers listing "union threat" as the primary motivation for their switch from hand to electronic sorting techniques was 730 acres. The median acreage grown by farmers listing "economic" and "personal management" as primary motivations for their change in harvest technology was 500 acres.

The electronic sorter not only generates some key benefits for growers ignored by the economic surplus approach. It also fails to generate some important costs to growers generally attributed to capital-intensive farm technologies in the literature on this issue. The literature on technological change in agriculture suggests that elimination of the small farmer often follows the adoption of a highly capital-intensive farming innovation. According to this literature, larger growers are usually the early adopters of a new technology. Smaller growers who are unable to afford the new technology quickly find themselves at a competitive disadvantage with their larger, more technologically advanced counterparts. Unable to produce as efficiently as their larger neighbors, smaller growers ultimately are forced to abandon farming the technologically affected crop. Accurate as this scenario may be regarding other technological innovation, it did not describe the effect of the electronic sorter on the small tomato grower in California during the years immediately following its adoption. This new technology clearly did not force a significant number of small growers out of tomato farming. In fact, during the first three years after the electronic sorter was introduced into the processing tomato industry, there was a 30 percent increase in the number of farmers growing processing tomatoes in California. Moreover, when small growers were interviewed in 1978, they did not express any fears about the adverse effects of the new technology or their ability to survive. Only 29 percent of the growers interviewed in Yolo and San Joaquin counties, for example, felt that adoption of the electronic sorter would affect them adversely. The vast majority thought it would have no effect whatsoever on them.

In light of the extensive literature linking the demise of the small farmer to more capital-intensive technological change, this is a particularly puzzling response. No doubt part of the explanation for it lies in the fact that the electronic sorter has not reduced production costs as dramatically as many other capital-intensive innovations in farming. The seventy-two cents per ton savings credited to the electronic sorter amounts to only a 2 percent reduction in growers' production costs. [36] This is a far cry from the savings generated by other more renowned technological advances in agriculture. The mechanical harvester itself, for example, reduced growers' production costs by 30 percent. [37] Because the savings generated by the electronic sorter are slight, smaller growers who might be unable to afford this latest technology are only at a slight competitive disadvantage vis-a-vis larger, more technologically advanced growers. As of 1978, this slight competitive disadvantage was not sufficient to force smaller growers out of the processing tomato industry. In part, this was because the successful collective price bargaining effort of the CTGA during these years sharply reduced the competitive pressures faced by tomato growers. This, in turn, permitted many smaller growers to remain in business even though they might have been slightly less efficient than their larger counterparts.

THE COSTS AND BENEFITS OF ELECTRONIC SORTING REEXAMINED

I began this chapter as an inquiry into the costs and benefits generated by adoption of electronic sorting techniques in the tomato industry. I wanted to learn who gained and who lost from adoption of this new technology. I also wanted to learn whether the total benefits generated by this new technology outweighed its total costs. In an effort to answer these questions, I first turned to the economic surplus approach to measuring costs and benefits commonly employed by agricultural economists. Using this approach, I estimated that, when fully adopted, the electronic sorter would generate an annual "gross social product" of $4,306,171. This gross social product results from a $10,311,346 gain in consumer surplus, and a $6,005,175 loss in producer surplus. I also estimated that the labor displacement costs generated by the new tomato harvest technology would lower its "net social benefit" anywhere from a high of $4,306,171 to a low of -$1,189,145 per year depending upon the reemployment possibilities of the tomato sorters affected by it.

While accepting estimates of the economic surplus approach as suggestive empirical referents, I felt that they were both incomplete and possibly misleading. They were incomplete because they failed to pin down more accurately the actual labor displacement costs that would be generated by the electronic sorter. They were possibly misleading because they were based on the assumed existence of competitive markets in the processing tomato industry. Concerned about the incomplete and possibly misleading estimates of the economic surplus approach, I felt it was necessary to take a "closer look" at the processing tomato industry in general and, more specifically, the main actors affected by the new tomato harvest technology. I believed this closer look would allow me to interpret the findings of the economic surplus approach in a more realistic light.

How did this closer look affect my evaluation of the electronic sorter? In general, I believe that it cast this new technology in a much less favorable light than did the starkly quantitative estimates of the economic surplus approach. First, the U.C. survey of tomato sorters strongly suggested that the vast majority of these workers will not find other employment as their work in the tomato harvest disappears. This meant that, at least in the short run, the labor displacement costs generated by the new technology might easily reduce its "net social benefit" to only $1,778,326 per year. Moreover, this same survey of tomato sorters indicated that farmworkers being displaced by the new tomato harvest technology already subsist at income levels significantly below those enjoyed by the average consumer or tomato grower. This, in turn, provided a reasonable justification for weighting the economic losses of farmworkers more heavily than the equivalent dollar losses or gains experienced by consumers and growers. I pointed out that even if we weighted the dollar losses of displaced farmworkers by only a factor of 1.7, they would cancel out the remaining "net social benefit" that might be credited to the electronic sorter.

I also took a closer look at tomato growers and tomato processors. My analysis of these two economic actors indicated that both are now or have been recently capable of exercising significant market power in their dealings with other actors in the tomato processing industry. This, in turn, strongly suggested that the $10,311,346 gain in consumer surplus credited to the electronic sorter by the economic surplus approach is a significant overstatement of these benefits. It even opened up the possibility that tomato growers and processors might be able to usurp all the benefits the new harvest technology generates, leaving only the costs of this innovation to other groups in society.

A close look at the electronic tomato sorter, the processing industry and the major groups affected by this new technology, indicated that this particular innovation could potentially generate more costs than benefits. This conclusion challenges the widely held view that virtually any further farm mechanization is bound to generate more good than evil. It also reaffirms the conclusion reached in the literature review conducted in chapter 2 that the gains and losses generated by technological changes in agriculture vary sharply from one socioeconomic setting to another. Ten years prior to adoption of the electronic sorter, the mechanical tomato harvester itself dramatically altered the socioeconomic setting into which the most recent tomato harvest technology would be introduced. Most importantly, the mechanical harvester sharply reduced the number of tomato growers in the industry. According to the records of the California Tomato Growers Association, there were over 4,000 tomato growers in California before the mechanical harvester was introduced in 1964. By the time the harvester had been fully adopted, however, there were less than 600 growers left in the state. Much of this concentration in the grower ranks could be directly traced to introduction of the new technology. As pointed out by Amy Barton and William Friedland:

> The machine required a sizable capital investment. The approximate $25,000 price of the early machines immediately began a process of concentration. Where a grower could, before, put in small acreages, it now became necessary to commit oneself to acreages approximating the capacity of the machine. Thus, with the early machine's capacity of about 75 acres, it made little sense to buy a machine if a grower was planting only 25 acres. [38]

By decreasing the number of farmers who were growing tomatoes, and by increasing significantly the capital investment of those who remained, the mechanical harvester served as a powerful stimulus towards grower organization. It was much easier to organize 600 larger growers than it had been to organize the 4,000 plus smaller growers of the pre-machine era. Moreover, once growers were forced to make the large fixed capital investment required by the mechanical harvester, they could no longer protect themselves from wide fluctuations in the market price of their produce. If the price of tomatoes fell sharply a year or two in succession, a grower in the machine era could lose everything because he would have no way of meeting his costly equipment payments. He was in a much more vulnerable position than a grower in the pre-machine era who, when faced by declining prices, could simply shift to another crop which promised better returns per acre. The only answer to this increased market vulnerability was grower organization for purposes of collective price bargaining with the food processors.

The processing tomato industry was by no means the model of a perfectly competitive industry even before the mechanical harvester was introduced. As early as 1956, eight of the major tomato processing firms in the industry accounted for more than 50 percent of the total production of processing tomatoes. But, by directly contributing to the organizational success of California tomato growers, the mechanical harvester pushed the industry even further away from the competitive market structure required to insure that the gains generated by subsequent technological changes would necessarily be passed on to consumers. Introduced as it was into the highly organized processing tomato industry, there was little chance that the electronic tomato sorter would generate the major consumer gains that it might have in a different, more competitive environment.

The traditional wisdom holds that whenever a new labor displacing technology is adopted in farming, it will generate serious short-term losses. But, as those displaced find alternate employment, or retire, the consumer benefits generated by the new technology ultimately will outweigh its short-term costs. This wisdom always has been unduly optimistic, in large part because it ignores the irreparable damage a displaced worker's impoverishment does to his children. But, when it is used to explain the consequences of mechanization in an organized sector of the economy like the processing tomato industry, the traditional wisdom becomes even more unrealistic. As I have pointed out earlier, the fact that some of the major actors in this industry exercise a significant degree of market power means that they may never have to pass on to consumers enough of the savings generated by the new technology to counterbalance the social costs of continued unemployment and poverty also caused by it. If the portion of these savings that are ultimately passed on to consumers fail to cover the losses suffered by displaced farmworkers and their children, then even in the long run the costs of the electronic tomato sorter may continue to outweigh its benefits. Whether this apparently real possibility will, in fact, materialize depends ultimately on government policies affecting everything from anti-trust activity to relief and retraining programs for the unemployed and impoverished.

Footnotes

1. Figures on the number of field hands and sorters involved in the harvest of processing tomatoes, as well as a useful short history of this industry are found in William H. Friedland and Amy Barton, Destalking the Wily Tomato, Research Monograph Number 15, Department of Applied Behavioral Sciences, University of California, Davis, June 1975, pp. 21-33, 41.
2. This figure was provided by the president of Blackwelder Farm Machinery Company in a personal interview during the spring of 1978.
3. Figures regarding the number of sorters employed on different kinds of harvesters are provided by Jon Brandt in An Economic Analysis of the Processing Tomato Industry, Unpublished Ph.D. dissertation, Department of Agricultural Economics, University of California, 1977, p. 72.
4. Melvin P. Zobel, Machine Harvest Cost--Processing Tomatoes--1977, Cooperative Extension Service, University of California, March 1978, p. 17.
5. The supply and demand schedules for processing tomatoes in California were estimated with a "pooled" econometric model developed by Wen S. Chern and Richard E. Just in Econometric Analysis of Supply Response and Demand for Processing Tomatoes in California, Giannini Foundation Monograph Number 37, September 1978, pp. 76, 77. Chern and Just presented several models which could be used to estimate the supply and demand response for tomatoes. My figures on the economic surpluses generated by the electronic sorter were developed using "estimation method I (OLS)."
6. The formulas used to calculate these costs and benefits are provided by Reed Hertford and Andrew Schmitz in "Measuring Economic Returns to Agricultural Research," Resource Allocation and Productivity in National and International Agricultural Research (Minneapolis: University of Minnesota Press, 1975), pp. 154, 155.
7. These costs are arrived at by multiplying the average hourly wage earned by tomato sorters in the nine years prior to the introduction of the electronic sorter ($1.84) by the number of hours of labor per acre displaced by the new technology (17.5) by the average annual acreage of tomatoes grown in the nine counties considered in this study during the pre-innovation period (170,662).
8. These surveys were funded as California Agricultural Experiment Project Number 3559. The questionnaires Ann Schuerring and I developed for these interviews are reproduced in Appendix A.

9. See, for example, Harold Sheppard et. al., Too Old To Work--Too Young To Retire: A Case Study of Permanent Plant Shutdown, Senate Special Committee on Unemployment Problems, 1959. Or, Sar Levitan and Harold Sheppard, "Technological Change and the Community," Adjusting to Technological Change, eds. Gerald G. Somers et. al. (New York: Harper and Row, Publishers, 1963), pp. 169-173.
10. Bureau of the Census, U.S. Department of Commerce, "Money Income and Poverty Status of Families and Persons in the United States," Current Population Reports Series, No. 116, 1977, p. 60.
11. Ibid.
12. Bureau of the Census, U.S. Department of Commerce, Current Population Report Series, No. 115, 1977, pp. 206, 207.
13. See, for example, Andrew Schmitz and David Seckler, "Mechanized Agriculture and Social Welfare: The Case of the Tomato Harvester," American Journal of Agricultural Economics, Vol. 52, November 1970, p. 574. Or, Theodore W. Schultz, "A Policy to Redistribute Losses From Economic Progress," Journal of Farm Economics, 59 (May 1977) p. 245.
14. David M. Gordon, Theories of Poverty and Unemployment: Orthodox, Radical, and Dual Labor Market Perspectives (Lexington, Mass.: D. C. Heath, 1972), pp. 1-172.
15. N. R. Collins et. al., Grower-Processor Integration: A Study of Vertical Integration Between Growers and Processors of Tomatoes in California, University of California, Agricultural Experiment Station Bulletin 768, 1959, p. 50.
16. Jon Brandt, op. cit., p. 30.
17. Collins et. al., op. cit., p. 57.
18. Jon Brandt, op. cit., p. 49.
19. Ibid., p. 30.
20. Ibid., p. 49.
21. Ibid., p. 30.
22. This figure was provided by Evan Husbeck, a spokesman for the CTGA during a personal interview in the spring of 1978.
23. Ehud Yonaz, "You Say Tomato, I Say Conspiracy," New West, September 25, 1978, p. 78.
24. Ibid., p. 79.
25. Ibid., p. 80.
26. Ibid., pp. 77-86.
27. Friedland and Barton, op. cit., pp. 30-31.
28. For a useful summary of this period in the industry, see Friedland and Barton, op. cit., p. 51.
29. This figure was provided by Evan Husbeck, a representative of the CTGA during a personal interview in the spring of 1978.
30. Ibid.

31. Sacramento Bee, July 21, 1981.
32. Ibid.
33. Jon Brandt, op. cit., p. 50.
34. Garden Valley Foods v. CTGA, Tri-Valley Growers and California Canners and Growers, California, 1979.
35. CTGA v. Tilley Lewis Foods, California, 1979. And, CTGA v. H. J. Heinz, California, 1980.
36. This figure is arrived at by dividing the estimated 72 cents per ton savings attributed to the electronic sorter by the total per ton cost of production for tomatoes in 1977 ($40.67).
37. This figure is arrived at by dividing the highest per ton estimate of costs saved by the mechanical tomato harvester ($7.47) by the total per ton cost of production for tomatoes in 1967 ($24.71).
38. Friedland and Barton, op. cit., p. 53.

5
Conclusion

I began this research as an inquiry into the question of who benefits and who loses from technological change in American agriculture. In an effort to shed some light on this issue, I first reviewed the existing literature on this subject. The major conclusion I drew from this literature was that there is no single political economy of mechanization in agriculture. The costs and benefits generated by new farm technologies varied sharply from one case to another depending upon: (1) the degree of competition in markets whick link farmers to consumers; (2) the degree of organization in the labor forces which handle farm products throughout the food chain; (3) the distribution of capital and other technological prerequisites among farmers in the affected commodity-industry; (4) the size of the stimulus the new farm technology provides for related non-farm industries; (5) the general vitality of the non-farm sector of the economy; and (6) the work related skills of displaced farmworkers. This conclusion seemed at very least to suggest the need for a case by case approach to mechanization in agriculture. It also lent a certain plausibility to the argument that in some cases the benefits of mechanization research may be usurped by narrow agribusiness interests at the expense of the wider public.

While a broadly based view of the literature gave plausibility to the argument that mechanization in agriculture might be benefiting the few at the expense of the many, quantitative studies of several different farm technologies indicated that this was highly unlikely. Using the economic surplus approach, these studies consistently indicated that new farm technologies generate a consumer surplus which far outweighs the costs these innovations impose on both farmworkers and other producers. This conclusion seemed to suggest that the concerns of groups opposed to mechanization research were unfounded after all. If consumers are the main beneficiaries of new farm technologies, then the gains of mechanization research clearly are not being usurped by a narrow range of agribusiness interests.

Because quantitative studies based on the economic surplus approach have been presented so often as convincing evidence in support of current State support for mechanization research, I decided that it was necessary to look at them more closely. After doing so, I concluded that this approach embodied certain assumptions and methodological choices which combined to provide an incomplete and possibly misleading view of the costs and benefits associated with new farm technologies.

In an effort to fill in some of the information gaps left by the economic surplus approach and to find out if studies employing this methodology were, in fact, presenting a misleading picture of the consequences of new farm technologies, I undertook a case study of the latest technological change in tomato farming. I wanted first to see what kind of a picture of the costs and benefits associated with the electronic tomato sorter would be provided by sole reliance on the economic surplus approach. I then wanted to see how this picture might be altered by a more empirical study of: (1) market conditions in the processing tomato industry; and (2) actual reemployment possibilities faced by displaced tomato sorters. As expected, the picture of the electronic sorter's costs and benefits indicated by sole reliance on the economic surplus estimates was a positive one. It indicated that, when fully adopted, the electronic sorter would generate a $10,311,346 gain in consumer surplus annually. It indicated further that even after discounting this gain in consumer surplus by the $6,005,175 loss the new technology would impose on producers, the electronic sorter would still generate a "gross social product" of $4,306,171 annually. Finally, while this approach did not provide an estimate of actual reemployment possibilities for displaced tomato sorters, it did indicate that even if only 22 percent of those displaced found new employment, the electronic sorter would generate a "net social benefit." I interpreted these results as providing a "favorable view" of the electronic sorter because they strongly suggest that consumers will reap major benefits from this new technology, and because they imply that in spite of its labor displacement effect, the sorter will likely generate a "net social benefit."

Unfortunately, a closer look at the processing tomato industry showed this "favorable view" of the electronic sorter to be extremely misleading. After surveying tomato growers and farmworkers, and after studying price and profit trends in the industry since the successful price bargaining efforts of the CTGA in 1974, I concluded that growers temporarily, and processors more permanently, had developed sufficient market power to alter the distribution of technological gains that the electronic sorter would have generated in a more competitive market setting. I also concluded that given the weakened competitive forces that existed in the processing tomato industy, it was impossible to estimate with any degree of certainty exactly how the gains generated by the new technology would be distributed. I noted, however, that it was very likely that the better organized growers and processors would be able to siphon off some and possibly all of the gains that would have gone to consumers in a more competitive industry. Finally, I pointed out that even if growers and processors were able to reduce consumer gains by only 20 percent, the probable unemployment generated by the latest tomato harvest technology would result in a "net social loss."

IMPLICATIONS FOR PUBLIC POLICY

What do the conclusions from the case study of the electronic tomato sorter and the review of the literature on technological change in agriculture suggest to government policymakers overseeing the State's role in technological change? First, they lend support to those voices which have been calling for a stop to the undiscriminating, "full speed ahead" approach to technological change that has been pursued by the USDA - Land Grant University Complex for the past three decades. If the costs and benefits generated by technological changes vary tremendously from one innovation to another and if, in some cases, the distribution of these costs and benefits may be disadvantageous to society, then surely government policymakers need to take a more discriminating look at the new technologies they support. Recent statements by former Secretary of Agriculture, Bob Bergland, indicate that at least some policymakers at the USDA are increasingly aware of this need. In a statement issued in January of 1980, Bergland noted that:

> I find it difficult to justify the use of federal funds to finance research leading to the development of machines or other technologies that may increase production and processing efficiency but at the same time damage the soil, pollute the environment, displace willing workers and reduce or eliminate competition. [1]

As this statement indicates, some policymakers at the Department of Agriculture are increasingly aware that there are significant social costs, as well as benefits, generated by new farm technologies and that there is a need to eliminate support for technologies that threaten a socially unacceptable set of costs and benefits. In particular, former Secretary Bergland has pointed the finger at research specifically aimed at reducing farm labor. According to Bergland, the USDA should not fund mechanization research where "other factors being equal or neutral, the major effect of that research will be the replacing of an adequate and willing work force with machines...." [2] This singling out of mechanization research for special scrutiny coincides with observations we noted in the literature review indicating that large-scale, capital-intensive technological changes have the greatest potential for generating a socially undesirable mix of costs and benefits. Unfortunately, however, it appears now as if the Reagan administration will quietly bury Bergland's suggestions regarding agricultural mechanization. [3]

Stirrings in the direction of a more discriminating agricultural research policy also have surfaced in state level policy-making institutions. In 1977, Assemblyman Torres introduced Assembly Bill No. 1192 before the California State legislature. This bill required the director of California's Food and Agriculture Department to complete a "social impact statement" before disbursing marketing order funds for research on "labor-displacing devices." As was stated in the bill itself:

> It is the purpose of this article to require the development of information on the results of labor-displacing devices prior to the commitment of funds for the research, development, or implementation of such devices so that a determination can be made of whether the proposed development is in the public interest prior to the expenditure of funds. [4]

Although Assembly Bill 1192 was never passed by the California legislature, its appearance on the "institutional agenda" clearly indicates a growing awareness on the part of some policymakers for a more discriminating agricultural research policy.

A primary argument against passage of the Torres Bill focused on its requirement that a social impact statement be completed <u>prior</u> to disbursement of any public research support for labor displacing farm technologies. Opponents of this legislation, including some who favored a more discriminating agricultural research policy, argued that it is impossible to anticipate accurately the sum total of social and economic effects of a new farm technology before it is even developed. [5] My own research lends support to this argument. Remember that upon concluding my review of the literature on mechanization, I noted that the costs and benefits generated in this process varied greatly from one innovation to another. Among the factors that contributed to this variation were the unique characteristics of the technologies involved. Since many of these characteristics cannot be known before the new technology is developed, any social impact study completed prior to the proposed technology development is likely to be very inaccurate.

While requiring social impact statements prior to development of new technologies seems unreasonable, it does seem sound public policy to mandate assessment of labor displacing farm technology at some stage before it's introduction in the market place. Once a new farm technology has reached a certain stage of development, a great many of its social and economic consequences can be accurately projected. A prime example of this kind of projection research has recently been provided by William Friedland et. al. for the already developed, but not yet adopted, mechanical lettuce harvester. Focusing their attention on both the structure of the lettuce industry and the characteristics of the lettuce harvester, Friedland et. al. were able to project the likely consequences of the harvester's adoption on labor displacement, the characteristics of the retained work force, work organization, grower-shipper concentration, housing, schools and attendant social services. [6] Of course, the accuracy of these projections ultimately must be judged against the actual effects of the mechanical lettuce harvester. There no doubt will be instances where these projections were in error. But, this is the fate of all predictive efforts in the social sciences. The effort to anticipate the consequences of a given action is the "raison d'etre" of social science. Why should this effort be suspended when it comes to predicting the consequences of technological change in agriculture?

If it is sound policy to assess the costs and benefits generated by labor displacing technological change in agriculture, it is also sound policy to regulate these changes in accordance with the discoveries of such assessments. Actually, the application of this belief has been under way for many years now with regard to most types of technological change. The Environmental Protection Agency, for example, already regulates a wide range of technologies in an effort to protect the natural environment.

Similarly, the Occupational Safety and Health Administration has broad authority to close down businesses utilizing production technologies endangering employee safety. It seems only logical to expand these regulatory efforts to deal with whatever additional social costs may be generated by new technologies. At present, one of the most damaging of these still unregulated social costs is labor displacement. Our study of the electronic tomato sorter, for example, suggested that certain capital-intensive harvest technologies <u>may</u> generate severe labor displacement problems without offering sufficient redeeming social benefits. More vigorous technology assessment efforts may help to identify these "anti-social" innovations. But, ultimately government policymakers must be willing to: (1) delay the adoption of these technologies until suitable social programs can be instituted to deal with their deleterious side effects; or (2) restrict the adoption of these technologies to periods of relative economic prosperity in hope of muting their most serious side effects, or in the most severe cases, prohibit the adoption of these technologies altogether. In short, government policymakers in this country must be willing to adopt the policy of "selective mechanization" that they have long been urging on decision makers in less developed nations. According to two well-known proponents of this policy, "selective mechanization means the type of mechanization that fits a country's economic, employment, production and cultural needs." [7] If selective mechanization makes sense for Third World nations, then it also should make sense in our own development.

A comprehensive policy of selective mechanization requires government policymakers to regulate labor displacing technology developed in both the public and private sectors. I noted above that some government leaders already have called for a more discriminating look at public research priorities. Others have demanded the preparation of social impact statements prior to disbursement of State funding for development of new farm technologies. To date, however, government leaders have shown no interest in regulating labor displacing technologies developed in the private sector. But, as we saw in the case of the electronic tomato sorter, these innovations can be as socially undesirable as those developed in the public sector.

No doubt part of the reluctance to regulate privately developed farm technologies stems from an unwillingness on the part of many government leaders to recognize the political character of so-called private economic decisions. As political scientists often have pointed out, the sole defining characteristic of a political decision is that it allocate authoritatively key societal values. From this perspective the choice of a more mechanized form of farming is a political decision regardless of who makes it. As professors Stout and Dowing have already pointed out:

> mechanization profoundly influences factors such as the volume and quality of production; the productivity of both land and labour; the cost of production; the level of employment; migration of agricultural wage earners and farmers; land ownership patterns; the development of mechanical skills and of a manufacturing and service related industry; and foreign exchange. [8]

By affecting such things as the level of employment, land ownership and foreign exchange rates, mechanization clearly serves to allocate key societal values. This in itself makes the choice of agricultural technology a political decision. Failure to accept this reality has prevented government leaders from formulating a more comprehensive and consistent regulatory policy toward technological change in agriculture.

I have suggested that government policymakers pursue both a more discriminating agricultural research policy and a more comprehensive regulatory policy with regard to labor-saving technological change. These measures alone, however, do not provide the necessary range of alternatives for a balanced public policy on technological change in agriculture. Many new farm technologies undoubtedly generate more benefits than costs. In these cases, the objective of government policy needs to be redistribution rather than regulation. Of course, government policymakers have long supported various programs ostensibly designed to redistribute gains generated by the economic system as a whole. But, technologically displaced workers seldom have been well compensated for the sacrifices they make for society. For example, total government transfer payments amounted to $224 billion in 1978. But, salaries and wages paid out in manpower training programs made up only $10 billion of this total. [9] I say "only $10 billion" because this seemingly large expenditure translates into only $125 per year for each of the approximately 8 million people who were unemployed that year. Even in the most generous states, unemployment insurance programs support most displaced workers at levels amounting to less than one-half of their previous earnings. Farmworkers, in particular, have been denied fair compensation for their job sacrifices in society's behalf. Until 1978, farmworkers in all but three states were blocked from collecting even the minimal Unemployment Insurance benefits available to most U.S. workers. Today only California, Hawaii, and Minnesota require all farmers to provide Unemployment Insurance to farmworkers. [10]

No doubt part of the explanation for the grossly inadequate levels of compensation currently provided to displaced workers can be found in the tendency of many policymakers to see any redistributive program as just so much more "welfare." In an effort to escape this pejorative labeling, some workers facing displacement have emphasized the desirability of explicitly linking unemployment and retraining compensations provided for workers to the technologies which have displaced them. Spokesmen for the UFW, for example, have suggested placing a small tax on all processed tomato products, the proceeds of which could be used to compenstate farmworkers displaced by the electronic sorter. Proponents of this perspective argue that it is only fair that major beneficiaries of a new technology also shoulder the major portion of social costs generated by it. They argue that if consumers are the real beneficiaries of such labor-saving farm technologies, then such a tax forces them to share a part of their gains with the farmworkers who are displaced by these same technologies. This approach to redistribution prevents the beneficiaries of a new farm technology from shifting the burdens of that technology to the general public. Also, by explicitly linking unemployment to the technologies that have generated it, public officials can more easily defend retraining and unemployment benefits as a worker's just dessert rather than a public handout.

The major drawback to source specific redistribution schemes, of course, are administrative. How would administrators decide, for example, exactly how much compensation technologically displaced farmworkers were entitled to? How would they decide exactly which farmworkers were entitled to this special compensation? Should it be available to any farmworkers who had ever harvested the affected farm commodity? Or, should it be limited only to those farmworkers who have harvested the affected commodity for a certain number of years? How many years must a worker be employed before he has a "right" to his job which justifies his being reimbursed for its loss? While admitting that questions like these pose thorny problems for administrators, advocates of source related worker compensation policies emphasize that this approach has already been used successfully in certain non-farm sectors of the economy. Since 1974, for example, American workers who have been laid off, or had their working hours cut due to increased imports, have been eligible for special federal benefits over and above normal unemployment compensations. These benefits which consist of job hunting help and cash payments of up to 70 percent of a worker's weekly wage are part of the Trade Adjustment Assistance Program. [11] This program is a conscious effort on the part of government policymakers to target a certain group of displaced workers for special benefits in return for losses suffered by them as a result of a planned effort to generate massive consumer benefits. A similar source related worker compensation program was recently set up by the federal government when it expanded California's Redwood National Park. In an effort to provide special compenation for the 300 lumber company employees who would be laid off due to the park's expansion, the federal government set aside $40 million to pay benefits to the displaced workers and to create new jobs for them. [12] Here again, the government decided that it was appropriate to provide special compensation to workers displaced by a political decision to generate benefits for another sector of society. The current administration is now making efforts to cut back on the amount of benefits provided displaced workers under both of the source specific redistribution programs mentioned here. But, they nonetheless continue to serve as working examples of programs that link compensation benefits for workers to the source of their displacement. Proponents of this approach conclude that if it can be used successfully to compensate labor displaced by foreign trade and recreational activities, it can be used to help labor displaced by agricultural mechanization.

Whether or not source specific redistribution schemes, or any of the other prescriptions I have outlined above, will be adopted depends, of course, on more than their administrative potential. It also depends to an important extent on the nature of the political climate surrounding technological change. Throughout this chapter I have prescribed more government regulation of new labor displacing technologies. I also have prescribed more government effort to redistribute the gains generated by these new technologies. These prescriptions seem reasonable, however, only if one sees the intrinsically political and therefore public character of technological choice. If one sees the State as an inevitable participant in technological choice, either by actively funding new technologies developed in the public sector or by passively legitimating technologies developed in the private sector, then it seems reasonable to restrict or prohibit innovations that are consistently prejudicial to certain social groups. But, those who see reality in this way apparently have taken on a minority status in recent years. Not long ago a solid majority of voters in this country elected a president who repeatedly stresses the private character of many technological and other decisions that are currently regulated by government. To the extent that the Reagan administration succeeds in promoting this perspective, my call for more government regulation of technological change may seem increasingly "unreasonable."

IMPLICATIONS FOR THE POLITICAL PROCESS

This research has focused primarily on the question of who benefits and who loses from mechanization in agriculture. By shedding light on the answer to this question, however, it also has contributed to the debate over the nature of the decision-making process behind these changes. This debate has grown out of two largely competing views of this process. I identified the first of these as the "pressure group" view. Its advocates have argued that, primarily because of the lobbying efforts of the agribusiness community, the public sector agricultural research complex has formulated its major problem as one of increasing production efficiency. Proponents of this perspective also have argued that the USDA - Land Grant University's indebtedness to agribusiness interests has led it to continue heavy support for mechanization research in spite of the serious costs this process imposes on other elements in society. Representatives of public sector research institutions, of course, have rejected this pressure group view of state supported agricultural research. Instead, they have offered a "public interest" view of the decision-making process behind agricultural research. Proponents of this perspective acknowledge that decision makers in the USDA - Land Grant University Complex have placed a heavy emphasis on mechanization research as a means of increasing production efficiency. They contend, however, that this strategy has not resulted from indebtedness to the agribusiness community, but rather from convincing evidence that the benefits generated by mechanization far outweigh whatever social costs this process may entail.

Which of these two competing views emerged from our analyses of the costs and benefits surrounding agricultural mechanization? In some ways, the answer is neither. The literature review conducted in chapter 2 indicated that mechanization in many cases may bring significant benefits to consumers. One can hardly infer from this conclusion that policymakers are promoting continued mechanizatin solely to advance the interests of agribusiness at the public's expense. On the other hand, the case study of the electronic sorter indicated that it is quite possible that in some cases agribusiness interests will capture the lion's share of technological benefits at the expense of farmworkers and consumers. In a world where well-organized economic actors bargain with each other over prices and profits, the benefits mechanization might generate for consumers may be usurped by growers co-ops, food processing oligopolies and labor unions. One can hardly infer from this observation that government policymakers continue to promote mechanization _only_ to advance the public interest. Apparently then, neither the pressure group nor public interest views portray a political decision-making process that completely squares with the distributive consequences of agricultural mechanization.

By itself, the inconclusive analysis of costs and benefits surrounding mechanization does not allow us to infer anything about the decision-making process behind this change. But, the continued vigorous promotion of this change by policymakers in the face of increasingly inconclusive evidence about its distributive effects may. It is not surprising that both farmers and farmworkers actively caught up in the mechanization process would adopt overly simplistic views of it to defend their economic self interests. It is surprising, however, that supposedly neutral, professional government administrators would continue to base public policy decisions on these gross oversimplifications. Our analysis of the electronic tomato sorter clearly indicated that the "evidence" used by public policymakers to defend their undiscriminating support of mechanization research was far from "convincing." As several critics have already indicated, it rests largely on an approach which assumes competitive markets exist throughout the food chain. That assumption obviously was not met in the processing tomato industry examined in this research. Nor is it likely to hold in many other commodity industries in American agriculture today.

The fact that government policymakers continue to promote an undiscriminating approach to mechanization can be traced to two possibilities. First, there is the increasing tendency of policymakers to rely on the summary judgments of experts in these matters. In the case of technological change, these experts have been largely social scientists who increasingly employ quantitative methods requiring overly simplistic assumptions about the real world. While those using such techniques often are aware of their methodological limitations, policymakers seldom are interested in the tenuous nature of results provided by various kinds of empirical research. They must make "defensible" decisions, and decisions based on "convincing evidence" are far more defensible than those based on "at best tenuous evidence."

There is, of course, a less indulgent inference we can draw about policymakers who continue to promote indiscriminate mechanization in the face of increasingly uncertain evidence regarding its distributive consequences. We can infer that their professional judgment about what constitutes "convincing evidence" has been clouded by their close ties with agribusiness interests. Agricultural policymakers themselves have acknowledged these close ties. They also have acknowledged that agribusiness interests regularly offer scientists and policymakers in the public sector opportunities for personal and professional gain. They have been careful to insist, however, that it is convincing evidence of mechanization's contribution to the public interest, and not agribusiness ties and inducement, which have led them to promote unrestricted advance of this process. My own research indicates that this cleansing rationale simply does not wash. This, in turn, strongly suggests that policymakers have acted in response to less public spirited motives.

The inferences one draws from the distributive effects of mechanization concerning the relative influence of public and agribusiness interests in the political decisions behind this process are necessarily tenuous. But, the inference one draws about the role of farmworkers in these decisions is unequivocal. Since its inception, mechanization has imposed heavy and obvious costs on displaced farmworkers. This fact has neither caused policymakers to slow the pace of this change, nor make any serious effort to cushion its blow. This could only have occurred in a political decision-making process where farmworkers were virtually powerless. The recent growth of the United Farm Workers Union has changed this situation somewhat. At least it has provided farmworkers sufficient political power to force the redistributive aspect of agricultural mechanization onto the public and institutional agenda of American politics. Whether this organization will acquire sufficient power to assure it meaningful participation in the policy-making process remains to be seen.

It clearly is inappropriate to infer much about the general character of the American policy-making process from a single case of technological change. But, the conclusions of this study of agricultural mechanization, and in particular the role of farmworkers in this process, do not square with those predicted by the pluralist perspective. Surely, the millions of farmworkers displaced by new farm technologies constitute a "legitimate social group" in

American society. Yet, their cries for social justice have been largely ignored by government policymakers for several decades.

Footnotes

1. Dallas Times Herald, February 15, 1980.
2. Ibid.
3. Sacramento Bee, March 9, 1981.
4. California, Assembly Bill Number 1192, Introduced by Assemblyman Torres, March 29, 1977, p. 2.
5. Sacramento Bee, February 17, 1978.
6. William H. Friedland et. al., Manufacturing Green Gold: The Conditions and Social Consequences of Lettuce Harvest Mechanization, Department of Applied Behavioral Sciences, California Agricultural Policy Seminar, University of California, Davis, July 1978, pp. 65-106.
7. B. A. Stout and C. M. Dowing, "Agricultural Mechanization Policy," International Labour Review, Vol. 113 (March-April 1976), p. 174.
8. Ibid., p. 176.
9. Lester C. Thurow, The Zero-Sum Society: Distribution and the Possibilities for Economic Change (New York: Basic Books, Inc., 1980), pp. 155, 160.
10. Stephen H. Sosnick, Hired Hands: Seasonal Farm Workers in the United States (Santa Barbara: McNally & Loftin, 1978), pp. 29, 269-278.
11. Los Angeles Times, April 18, 1977.
12. Wall Street Journal, March 29, 1978.

Appendix

FARMWORKER QUESTIONNAIRE

A. Work Experience as a Sorter.

1. How many seasons have you worked as a sorter on the tomato harvester?

2. How many weeks have you worked as a sorter this year?

3. How many days, on the average, have you worked during each of these weeks?

4. How many hours per day, on the average, have you worked on each of these days?

5. How much per hour are sorters on your harvester paid for their work?

6. Do they receive any bonus in addition to this hourly wage?

IF THE RESPONDENT ANSWERS YES, THEN ASK:

7. What kind of bonus do they receive?

8. What benefits, in addition to the wages or bonus you have already mentioned, are available to sorters working with you on the harvester?

IF THE RESPONDENT INDICATES ANY, THEN ASK:

9. Do you personally receive any of these benefits?

NOW, I WOULD LIKE TO ASK YOU A FEW QUESTIONS ABOUT YOUR WORK AS A SORTER LAST YEAR.

10. How many weeks did you work as a sorter last year?

11. How many days per week, on the average, did you work last year on the harvester?

12. How many hours per day, on the average, did you work last year?

13. How much per hour were sorters on your crew paid last year for their work?

14. Did they receive any bonus in addition to this hourly wage?

IF THE RESPONDENT ANSWERS YES, THEN ASK:

15. How much of a bonus did they receive?

16. Have you worked on harvesters which had electronic sorters, or only on machines where the sorting is done by hand, or both types?

17. Which of these types do you prefer to work on?

IF THE RESPONDENT EXPRESSES PREFERENCES, THEN ASK:

18. Why do you prefer this kind of machine?

19. What has been the average number of sorters working when you have been on:
___ harvesters with electronic sorters?
___ harvesters with manual sorters?

20. What aspect of your work on the harvester do you find most difficult?

21. Is there some aspect of this work which you find particularly enjoyable?

22. Have any other members of your family worked as sorters this season? By family, I am referring to any <u>relative</u> that lives with you most of the year.

IF THE RESPONDENT ANSWERS YES, THEN ASK:

23. How many?

24. Who? (What relation?)

25. Would you say that, on the average, they worked:
___ about as many hours as you?
___ about half as many hours as you?
___ less than half as many hours as you?

26. If you had not worked as a sorter this season, is there other work that you could have done?

IF THE RESPONDENT ANSWERS YES, THEN ASK:

27. What kind of work would that have been?

28. Do you personally know anyone who looked for work as a sorter this year, but was unable to find it?

29. Do you plan to look for work as a sorter again next year?

30. How confident are you that you will be able to find work as a sorter next year?
___ very confident
___ fairly confident
___ not very confident

TO THIS POINT MY QUESTIONS HAVE DEALT WITH YOUR EXPERIENCE WORKING IN TOMATOES. THE FOLLOWING SERIES OF QUESTIONS DEAL WITH YOUR WORK EXPERIENCES GENERALLY DURING THE LAST FEW YEARS.

B. General Work Experience

31. What do you consider to be your major occupation?

32. What has been your most important source of income this year? By this year I mean from January 1, 1977 to the present.

33. Who was your employer on this job?

34. Approximately, how many days per week did you work on this job?

35. Approximately, how many hours per day did you work on this job?

36. Approximately, how many weeks or months did you work on this job?

37. What was your rate of pay per hour or per week on this job?

38. What other jobs have you performed during this year?

REPEAT QUESTIONS 33 THROUGH 38 RECORDING THE APPROPRIATE INFORMATION ON THE RESPONDENT'S WORK HISTORY FORM FOR 1977 THAT APPEAR AT THE END OF THIS INTERVIEW.

REPEAT QUESTIONS 32 THROUGH 38 AS MANY TIMES AS NECESSARY TO EXHAUST THE RESPONDENT'S WORK HISTORY FORM FOR 1976.

39. What do you estimate your total income for the past year to be? Here I mean from July 1, 1976 to the present.

C. Work Preferences

40. If you could make a decent income in either agricultural work or non-agricultural work, would you prefer to:
___ remain in farm work?
___ find employment off the farm?

41. Why do you feel this preference?

42. Would you prefer that the other members of your family:
___ remain in farm work?
___ find employment off the farm?

43. Do you prefer seasonal work or would you rather have a year-round job?

44. Why?

WE HAVE NOW COMPLETED THE MAJORITY OF THE INTERVIEW, MR./MRS. __________. BUT, I WOULD STILL LIKE TO ASK YOU A FEW MORE QUESTIONS ABOUT YOU AND YOUR FAMILY. IF YOU FEEL THAT ANY OF THE QUESTIONS ARE TOO PERSONAL, PLEASE SAY SO AND I WILL SIMPLY GO ON TO THE NEXT ONE.

45. When were you born?

46. What was the highest grade that you completed in school?

47. Do you have any training or experience in work other than farm work? What?

48. Have you lived anywhere other than in Yolo/San Joaquin County during the past year?

IF THE RESPONDENT ANSWERS YES, THEN ASK:

49. Where? How long?

50. Are you single, married, separated or divorced?

51. Most of the time, do you live alone or with family? By "most of the time" I mean six months or more out of the year.

IF THE RESPONDENT LIVES WITH FAMILY, THEN ASK:

52. How many children are in the family? By children I mean individuals less than 18 years of age.

53. How many adults, other than yourself, are in the family? By adults I mean only those people who are 18 years or older.

IF OTHER ADULTS ARE IN THE FAMILY, THEN ASK:

54. Have the other adults in the family been employed any time during the past year?

IF YES, THEN ASK:

55. How many?

56. Which one of these people contributes the greatest amount of financial support to the family?

57. What has been his/her most important source of income this year? Again, by this year I mean from the 1st of January to the present.

USING QUESTIONS 32 THROUGH 38 AS A GUIDE, COMPLETE A WORK HISTORY FORM FOR THIS PERSON FOR BOTH 1977 AND 1976.

58. What do you estimate the income for your entire family will be in 1977?
___ less than $1,000
___ between $1,000-$2,000
___ between $2,000-$3,000
___ between $3,000-$4,000
___ between $4,000-$5,000
___ between $5,000-$7,000
___ between $7,000-$10,000
___ between $10,000-$13,000
___ more than $13,000

59. What do you estimate the income for your entire family was in 1976?
___ less than $1,000
___ between $1,000-$2,000
___ between $2,000-$3,000
___ between $3,000-$4,000
___ between $4,000-$5,000
___ between $5,000-$7,000
___ between $7,000-$10,000
___ between $10,000-$13,000
___ more than $13,000

60. Have you or other members of your family received any benefits from:
___ unemployment?
___ food stamps?
___ medical?
___ welfare?
___ any other program?

IF THE RESPONDENT INDICATES HE/SHE HAS RECEIVED BENEFITS FROM ANY OF THESE PROGRAMS, THEN ASK:

61. For how long did you receive these benefits?

62. How much did you receive in benefits from this/these programs?

63. What state or country do you consider your permanent home?

WORK HISTORY 1977 (RESPONDENT)

Job Title	Employer	Hrs./day, days/week, weeks/year	Earnings

WORK HISTORY 1976 (RESPONDENT)

Job Title	Employer	Hrs./day, days/week, weeks/year	Earnings

GROWER INTERVIEW

I. Background of Grower

1. What is your present position on this farm, (Name of Interviewee)?

2. How many years have you been in farming?

3. How many of those years have you been in your present position?

4. What other occupations have you held in agriculture?

5. Do you live on the farm?

IF THE RESPONDENT ANSWERS NO, ASK THE FOLLOWING QUESTION:

6. Approximately how far do you live from the farm?

7. What is the highest grade you completed in school?
___ less than a high school graduate
___ high school graduate
___ some college
___ college graduate
___ some graduate work
___ advanced degree

IF THE RESPONDENT WENT TO COLLEGE, ASK THE FOLLOWING QUESTIONS:

8. What college did you attend?

9. What was your major field of study?

10. What is your ethnic background?
___ Asian American
___ Black
___ Mexican American
___ Native American
___ Other Caucasian

11. Were you born in the U.S.? IF NO, where?

II. Type of Farming Operation

12. How many years have you been growing tomatoes?

13. Approximately how many acres of tomatoes have you grown in each of the past three years? 1977? 1976? 1975?

14. What was your average yield per acre in each of these years? 1977? 1976? 1975?

15. Again, for each of the last three years, what percent of your tomatoes have been grown on land that you own and what percent have been grown on land that you lease? 1977? 1976? 1975?

16. How many regular employees do you hire on this farm? By regular I mean employees who worked here at least five months per year.

17. How many regular employees did you hire in each of the past two years? 1976? 1975?

18. What has been the average number of seasonal workers you have hired for each of the past three tomato harvests? 1977? 1976? 1975?

19. How do you usually recruit the majority of your seasonal workers?
___ direct personal contact
___ E.D.D.
___ labor contractor

20. How do you usually supervise your seasonal workers?
___ direct personal contact
___ through a foreman/supervisor

21. Where do you think the majority of your seasonal workers live? By seasonal workers I am referring now only to those that you hire during the tomato harvest.
___ in housing provided by you on the farm
___ in publicly supported Farm Labor camps
___ in their own homes locally

22. Do you feel that the majority of your seasonal workers this year are locals or migrants?

23. Have you been short of seasonal workers for the tomato harvest this year? Or, do you feel there were an excess of workers this year?

24. Have you been short of seasonal workers for the tomato harvest during any of the past few years? IF YES, which years?

25. Generally speaking, how confident are you that there will be enough seasonal workers available to harvest your tomatoes in the next few years?
___ very confident
___ fairly confident
___ not very confident

26. Why do you feel this confidence/lack of confidence?

27. What has been your average turnover among employees who are hired only for the tomato season?
___ less than 25%
___ between 25% and 50%
___ greater than 50%

28. What other crops do you grow on this farm? Number of acres?

29. Of these, which also demand extra workers at peak times?

30. How would you prefer to recruit field workers?
___ labor contractor
___ E.D.D.
___ union
___ direct contact with workers

30a. Generally speaking, do you believe a local labor pool could be developed capable of handling your farm labor needs?

IF THE RESPONDENT ANSWERS NO, PROCEED TO PART III.

30b. Maintaining such a pool of reliable local labor has certain costs. Do you think growers would be willing to increase the wages presently paid to farm labor to accomplish this objective?

III. Mechanization: Motivations for Adoption and General Experiences.

31. How many mechanical tomato harvesters will you be using this year to bring in your crop this season?

32. How many of these are owned by you, and how many are only leased or contracted by you?

33. Can you give me this same information for each of the past two years? 1976? 1975?

34. How many of these mechanical harvesters have used the electronic sorter? 1977? 1976? 1975?

35. What has been the average number of crew members you have had working on the harvester this year?

36. What sort or workers do you prefer to hire as sorters?
___ migrants
___ local laborers
___ women
___ men
___ younger workers
___ older workers (over 25)

IF THE RESPONDENT HAS NO ELECTRONIC SORTERS, PROCEED TO QUESTION #51.

37. What were the major factors that you considered when you thought about investing in an electronic sorter?

38. What was the most important factor that influenced you to buy a sorter?

39. Did you do a cost analysis before investing in the sorter?

IF YES, THEN ASK:

40. How much did you estimate it would reduce your harvesting costs in dollars/tons?

41. How many years do you estimate it will take to amortize the cost of the electronic sorter?

42. If the electronic sorter becomes the primary means of harvesting tomatoes, will it affect your competitive position in relation to other growers?

IF THE RESPONDENT INDICATES ADVANTAGE/DISADVANTAGE, THEN ASK:

43. Why will you be affected favorably/unfavorably?

44. Have you been able to use the electronic sorter reliably?

45. Under what conditions can it not be used?

46. Have the machines with electronic sorters required more maintenance than those without them?

47. Is there more off-the-farm maintenance required by the electronic sorter than by harvesters without them?

48. What percentage of your crop is left in the field when you harvest with an electronic sorter?

49. Is this more, less or about the same waste of tomatoes that you get using harvesters with manual sorters?

50. Has the sorter proved satisfactory in cutting costs? Adding reliability to grade? Speeding the harvest?

51. Do you plan to use (additional) electronic sorters next year? IF YES, how many?

52. Do you (would you) prefer harvesting tomatoes with an electronic sorter?

53. If the costs in both time and money were exactly equal, would you prefer to harvest tomatoes with an electronic sorter or with field laborers?

IF THE RESPONDENT STATES PREFERENCE, ASK WHY?

54. Do you believe the electronic sorter will be universally adopted in the near future?

55. Is there any circumstances that might slow this adoption? IF YES, what?

IV. Social Consequences

56. Who, if anyone, do you think should be responsible for finding new employment for those workers whose jobs are eliminated due to advances in farm technology?

WELL, MR./MRS.__________, THAT CONCLUDES OUR QUESTIONS. IS THERE ANYTHING YOU THINK IS IMPORTANT THAT YOU MIGHT LIKE TO MENTION TO US?

For Product Safety Concerns and Information please contact our EU representative GPSR@taylorandfrancis.com Taylor & Francis Verlag GmbH, Kaufingerstraße 24, 80331 München, Germany

Batch number: 10725937

Printed by Printforce, the Netherlands